CONSTRUCTION DETAIL BANKING

CONSTRUCTION DETAIL BANKING

Systematic Storage and Retrieval

PHILIP M. BENNETT, R.A.
Architect and Professor

A Wiley-Interscience Publication

JOHN WILEY & SONS

New York · Chichester · Brisbane · Toronto · Singapore

FRONT COVER

CAD workstation is courtesy of the
Summagraphics Corporation.

DETAIL QUALITY

The details presented in this manual have been selected to
demonstrate graphic concepts rather than technical quality.
They do not necessarily represent current solutions to
present-day design problems. Their primary function is to
show how design information can be standardized in the
production of working drawing details.

Library of Congress Cataloging in Publication Data:

Bennett, Philip M., 1936-
  Construction detail banking.

  "A Wiley-Interscience publication."
  Bibliography: p. 169
  Includes index.
    1. Information storage and retrieval systems—
Building.   2. Building—Details—Data processing.
I. Title.
Z699.5.B8B46 1984   025'.04   83-14490
ISBN 0-471-88621-1

Printed in the United States of America

10 9 8 7 6 5 4 3 2 1

PREFACE

The technology explosion of the 1900s has generated large volumes of valuable construction information that have been ineffectively accessed by design professionals. Fragmented attempts to develop information-handling systems have created negative reactions to using data bases. Many information storage and retrieval programs have been individually structured around the requirements of limited subject areas and small user groups. As a result the design professional is still left with unstructured formats for developing in-house information systems.

This manual sets forth guidelines and procedures for developing an information-handling program for working drawing details, specifications, and construction information. It is structured for architects, engineers, landscape architects, interior designers, industrial designers, and all others who prepare construction drawing details. Design educators will also find this system appropriate for teaching information-handling techniques in their curriculum.

The ultimate purpose of the information-handling principles generated in this manual is to advance national and in-house design information storage and retrieval systems. The versatility of these principles will provide maximum flexibility for processing a diverse range of construction subjects based on the communication language of the user group. Designers communicating through a medium of working drawings and details will gain valuable concepts for structuring in-house information-handling programs. By following the recommendations and procedures set forth, design professionals can expect to:

1. Develop effective detail storage and retrieval systems.
2. Maximize use of evaluated design data.
3. Improve the quality of new information.
4. Reduce time spent on repetitive activities.
5. Access information quickly and efficiently.
6. Develop uniformity among all construction documentation.
7. Decrease errors in construction communication.

The insights gained to develop this manual were derived from my firsthand experience in coordinating the environmental design information program for the ERIC Clearinghouse on Educational Facilities. This experience combined with a national and international research study on the effectiveness of construction information storage and retrieval has helped to crystalize ideas and concepts for structuring a "master or universal" information system. Every effort has been made to set forth a system based on the positive learning experiences from many facets of information handling rather than reinventing a new system for each user group.

Many valuable experiences in processing design information have been identified to provide a comprehensive overview of the essential elements for construction detail banking. Twelve years of developing continuing education programs on Working Drawing Production Techniques at the University of Wisconsin have helped to define clearly a need to establish a constructive framework for information handling. My exposure to many design professionals with questions and problems concerning graphic communication has also helped to formulate the recommendations presented in the manual.

I thank many individuals and firms for willingly contributing support information to reinforce specific concepts and procedures identified in this manual. (Material taken from other sources is identified by a source note and is followed by a graphic symbol, □, identifying the end of the sourced material.) Their valuable examples have helped present a system for developing and processing reusable construction information. Each firm's positive and negative learning experiences have further helped to formulate guidelines for effective detail banking. I extend special thanks to my son Ryan who assisted me in proofing and editing this manual.

PHILIP M. BENNETT

Sun Prairie, Wisconsin
February 1984

CONTENTS

Chapter 1   Introduction   1

    1-1   Purpose, 1
    1-2   User Group, 2
    1-3   How Can This Manual Be Used? 2

Chapter 2   Selection of Standard Details   3

    2-1   Importance of Good Details, 3
    2-2   Criteria for Selecting Details, 5
    2-3   How Do You Begin Detail Banking? 7
    2-4   Developing a Detail Grading Sheet, 9
    2-5   Detail Selection Procedures, 9
    2-6   Manufacturers' Standard Details, 15
    2-7   Photographic Details, 15
    2-8   Association Details, 15
    2-9   Architectural and Engineering Details, 18

Chapter 3   Preparation of Standard Details   23

    3-1   Importance of Quality Control, 23
    3-2   Planning for Detail Production, 23
    3-3   The Detail Development Process, 25
    3-4   Standard Detail Requirements, 26
    3-5   Quality Control During Detail Development, 47
    3-6   Development Quality Control Procedures, 48
    3-7   Field Evaluation Procedures, 52
    3-8   Development of a Field Evaluation Data Card, 55
    3-9   Benefits of a Field Evaluation Program, 57
   3-10   Developing a Detail Data Card, 59
   3-11   Detail Development and Use Cycle, 61

Chapter 4   Development and Use of a Construction Language   63

    4-1   Purpose, 63
    4-2   Problems in Handling Information, 64
    4-3   Support for a Construction Language, 68
    4-4   Procedures for Selecting and Developing a National
           or In-house Language for Detail Banking, 73
    4-5   Structuring a Thesaurus of Construction
           Terminology, 81

4-6   Directives for Creating a Language and a
        Thesaurus, 90

Chapter 5   Evaluating Information Handling Systems   **93**

5-1   Banking System Requirements, 93
5-2   Types of Indexing Systems, 94
5-3   How to Evaluate a System, 112

**Chapter 6   Detail Banking Procedures: How to Develop a
                    Master System**   **115**

6-1   How to Develop the Descriptor Term
        System, 115
6-2   Critical Steps in Detail Banking, 119
6-3   Specifications Storage and Simultaneous Retrieval
        with Details, 129
6-4   Methods of Storing and Retrieving Details, 133
6-5   How to Develop Numbering Systems, 142
6-6   Evaluating Methods of Indexing and Storing
        Details, 147
6-7   Effective Methods of Searching and Retrieving
        Details, 151
6-8   Why Choose the Recommended Master
        System, 154

Chapter 7   Developing a Comprehensive Information System   **157**

7-1   How to Develop a Comprehensive Information
        System, 157
7-2   Keeping Pace with a Changing Communication
        Technology, 159
7-3   Planning for Change, 160
7-4   Developing a Network of Information
        Exchange, 166

Glossary   **167**

References   **169**

Further Reading   **171**

Index   **173**

CONSTRUCTION
DETAIL BANKING

ONE
INTRODUCTION

1-1 PURPOSE

Within the design field, technological change and a rapid increase in information have surpassed development and organization of effective information-handling systems. A lack of direction in handling information has caused many design professionals to develop individual systems. These systems have helped create fragmentation and confusion in processing design information. Most of the existing systems are unique and contain symbolism that confuses and discourages many potential users. This manual has been developed to organize a communication system that is applicable to the entire construction industry.

The system set forth in this manual provides guidelines and procedures to develop an information-handling system for working drawing details, specifications, and construction information. This system is based on an overall communication program that will enable design professionals to handle effectively all information generated within the design office. The working drawing data bank can be considered one division of this overall information-handling system. The framework for this system uses terminology in the English language. Terminology control has been further established by the Joint Council of Engineers. A historical base has been developed by similar systems in use by the federal government, private industry, and foreign countries. The concept for this system has also gained support of the construction industry in both Canada and the United States.

This system is not to be considered an individual concept, but a program based on a sound communication language with guidelines set forth by a recognized body of professionals. It differs markedly from many existing systems, in that

segments of it have been utilized effectively by a number of discipline areas. The main advantages are flexibility and open-ended expansion for variations in information handling required by design professionals.

Standardization measures and quality control procedures are identified to assist the production staff in preparing the best information for construction communication. Therefore construction terminology used in graphic information constitutes the language required for detail banking. A master system is developed to index, store, and retrieve all construction details selected for banking systematically.

The objectives of detail banking can be summarized as follows:

▷ To establish a master system for storing and retrieving construction details.

▷ To encourage standardization and storage of reusable construction information.

▷ To encourage quality control and field feedback programs to improve working drawing production.

▷ To develop a storage and retrieval system that is usable by all disciplines in the construction industry.

▷ To build a communication language that unifies all construction documents and establishes the basis for storing and retrieving construction information.

▷ To develop a storage and retrieval system that is language oriented with total flexibility to change with technology.

▷ To reduce, if not eliminate, the need for creating unique "individual" storage and retrieval systems that are difficult to comprehend and have limited flexibility for use.

1

1-2 USER GROUP

All disciplines in the construction industry that communicate with detail drawings can use this document to increase their effectiveness in information handling. The language-based information system described makes it possible for many disciplines to use this concept by simply generating the terminology characteristic of their profession. System development guidelines have been structured for:

▷ Architects
▷ Engineers
    Mechanical
    Electrical
    Civil
    Structural
▷ Landscape Architects
▷ Interior Designers
▷ Industrial Designers
▷ Design Educators

Architectural and engineering schools will find this manual appropriate for teaching information-handling techniques in their curriculum. Design professionals in the United States, Canada, and several European countries presently working with a standardized construction language will also find this manual a valuable resource.

The recommended "master information system" can be easily developed and implemented by design professionals who establish a standard vocabulary derived from their communication language. Basic procedures for developing and using this system will remain constant for all disciplines. These procedures will guide each user group in structuring a construction language and thesaurus appropriate for information handling. A standard construction language will facilitate detail banking and information handling for all disciplines within the construction industry.

1-3 HOW CAN THIS MANUAL BE USED?

The master system presented in this manual is structured to guide the design professional in creating an information storage and retrieval system for handling working drawing details. Specific guidelines and procedures are set forth to develop and administer an effective detail data bank. The structure for this system is sufficiently broad and flexible to encompass development of an overall information-handling program within the design firm. The detail data bank treated in this manual will be only one segment of the overall system for information processing by the design professional.

In proceeding to develop the system, it is important to research the technology level of communication equipment. This will enable the user to design an information system for most effective use of equipment. The guidelines set forth in this book will provide design professionals at all levels with manual, semiautomated, and automated programs to structure procedures for effective storage and retrieval of working drawing details. Whatever level is selected, these guidelines will serve as the basic principles and procedures to be followed while advancing the system into higher and more complex communication system technology. This system is open-ended and flexible and will allow an indefinite number of details to be handled over a long period of a firm's historical development.

TWO
SELECTION OF
STANDARD DETAILS

2-1 IMPORTANCE OF GOOD DETAILS

Successes and failures of today's buildings are to a large extent the result of today's working drawings. An increasing number of projects have major failures in less than one or two years after construction. Recent insurance studies indicate that as much as 65 percent of legal actions are due to problems in design and specifications, or the basic manner of construction communication [1].

Good, clear communicative details save time, money, and negative feedback. Effective details also avoid confusion that can result in construction errors. The test of a good detail is in field construction where graphic information must be interpreted to assemble building materials. A detail should be analyzed and evaluated before preparing working drawings. Figure 2-1 highlights the major advantages of developing standard details.

In today's construction environment it is almost impossible to maintain consistency in good detail development because of time and economic pressures. Several areas of concern must be taken into account before developing construction details.

2-1.1 Investigate and analyze the process of developing construction details.

Before starting on a detail banking system, it is important to study and identify successful procedures for creating effective details. Quality review of firms' past procedures will help to identify areas of needed improvement. The success of a banking system is dependent on communication from all individuals responsible for construction detail. Unless a system has been well organized, it cannot provide a proper flow of critical construction information. Each team member plays a valuable role in creating a successful standard detail. Once the process for developing standard details is set forth, the firm can proceed to structure an effective detail banking system.

2-1.2 Identify research and testing centers to obtain detail development information.

A search of information sources such as testing agencies and research organizations is critical in setting up the detail development and review process. During development of standard details, it is important to question material and construction relationships that may lead to future problems. Maintaining a file on information research centers will enable the design professional to access information quickly during the detail development process. A listing of resources and centers should be made a part of the office procedures manual for detail development.

2-1.3 Evaluate methods of storing and retrieving information from files.

Many different storage and retrieval programs exist on a national basis. Some of these programs work successfully for particular discipline areas of information. It is incumbent upon the design professional to evaluate those closely related systems to determine their effectiveness and po-

▷ **REDUCE RISK**

   —**Builds in field evlauation process**

▷ **REDUCE TIME**

   —**Development and production time savings by reducing repetition**

▷ **REDUCE CONFUSION**

   —**Detail clarity and simplification produces ease of interpretation during bidding**

   and construction

▷ **REDUCE COST**

   —**Substantial time savings produces cost savings**

▷ **REDUCE ERRORS**

   —**Creates an opportunity to evaluate materials and work out problem areas**

▷ **REDUCE UNKNOWNS**

   —**Provides an early prediction of the scope of new projects and aids preparation of**

   required details

Figure 2-1   Major Advantages of Standard Details.

tential for use in the working drawing field. Technology is also changing mechanical systems for storing and retrieving information. This equipment should also be evaluated for its potential use with the selected storage and retrieval program. Determining current information handling practices will enable the design professional better to select appropriate tools and procedures for developing a detail data bank. Review of private and governmental programs should be a critical step in developing a state-of-the-art study on information processing in the design field.

2-1.4 Study and develop field feedback procedures.

Success of any standard detail banking system depends on the quality of each detail that is entered. Unless a detail has had thorough study and field evaluation, it remains questionable whether or not the detail should be entered into the system. Therefore it is important to develop a good field feedback program for evaluating each detail used on the working drawings. Each firm anticipating development of a detail banking system should analyze its field feedback process to determine whether the flow of information is effective enough to properly evaluate each detail. Establishing a good quality control program is essential in structuring a successful detail banking system.

2-1.5 Establish a systematic process for research, testing, and developing details.

Success of any program depends on the procedures and systematic steps taken to fulfill the goals and objectives set forth in the program. Each area of study and review must be associated with a set of steps and guidelines for achieving that goal. Methods for determining the level of achievement must also be established in order to determine the success of the system. The word "systematic" is the key to the organization and administration of any effective detail banking program. The overall office policy and guidelines should allow the person in charge of the detail banking system to administer effectively a systematic process for achieving all goals established in each segment of the operation.

2-1.6 Establish communication lines with technical people in manufacturing and construction.

Contributors to a successful detail development program are the design team, manufacturing representatives, and construction team. Communication must take place between all team members to ensure successful development of a specific detail. Methods and techniques for achieving communication must be set forth early in the design development process to incorporate valuable construction experience. It is essential that a successful communication system be developed early in the detail banking program. Timing for detail reviews and research must be specified in order to have proper evaluation during each

stage of detail development. Construction techniques, materials, and detail relationships should all be checked and graded by each discipline represented on the design team.

An increasing number of legal problems associated with construction has led the way to more standardized details and greater field evaluation. Effective quality control programs have been instituted to maintain surveillance over all stages of detail development. The designer is compelled to become more concerned with every detail in a given building while keeping in mind the total end product. Each team member will also be required to do better research and acquire more field performance information in order to improve future building details.

The future indicates the design professionals will need to be greatly concerned about graphic communication for each construction detail. Documented research and material testing must be evaluated before construction details are produced. Field testing of details is necessary to gain appropriate feedback for future design. New information banks will systematically accommodate development of working drawing details while maintaining active files for storage and retrieval of evaluation data. Innovative methods of storing and retrieving details must be developed to alleviate increased pressure on the designer to utilize an unstructured information base

2-2 CRITERIA FOR SELECTING DETAILS

Design firms contemplating development of a detail data bank should place first priority on identification of selection criteria. It is important that specific goals and objectives be set forth before processing details, since each firm's needs differ. Each detail in the historic file has potential for reuse. Therefore selection criteria should be set forth that represent the firm's unique needs for developing construction details. No detail should be entered in a data bank without first being screened and evaluated against a set of selection criteria. It is also important to establish an evalua-

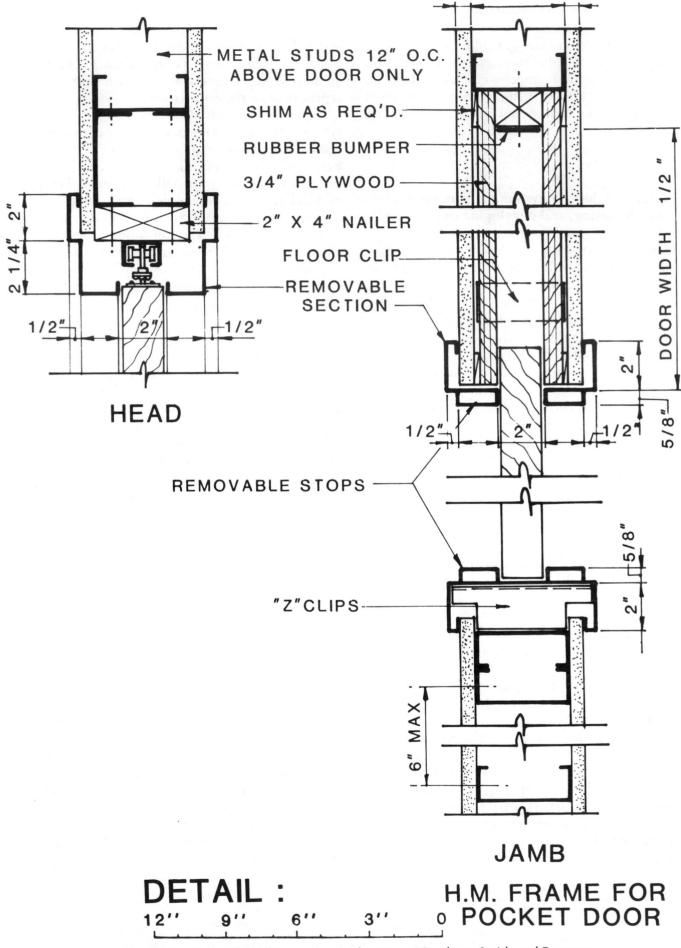

METAL STUDS 12" O.C. ABOVE DOOR ONLY

SHIM AS REQ'D.

RUBBER BUMPER

3/4" PLYWOOD

2" X 4" NAILER

FLOOR CLIP

REMOVABLE SECTION

2 1/4" 2"

1/2" 2" 1/2"

HEAD

DOOR WIDTH 1/2"

2"

5/8"

1/2" 2" 1/2"

REMOVABLE STOPS

5/8"

2"

"Z" CLIPS

6" MAX

JAMB

DETAIL :

12" 9" 6" 3" 0

H.M. FRAME FOR POCKET DOOR

Figure 2-2 Hollow Metal Frame for Pocket Door (Gresham, Smith and Partners, Nashville, Tennessee).

tion and approval process that grades each detail prior to entering the data bank.

Detail banking can help a firm improve overall quality of reusable or standardized details. This process gives the design professional an opportunity to select and reevaluate each detail as it is prepared for storage. Several criteria should be kept in mind while making the selection of reusable details:

1. Frequency of detail within a given structure.
2. Number of similar structures designed over a given time frame.
3. Impact of user requirements that cause details to change.
4. Rate of technology change that impacts on detail.
5. Percent of change required to reuse detail.
6. Is detail considered a standard in construction industry?
7. Does detail design represent a universal solution to problem?

2-3 HOW DO YOU BEGIN DETAIL BANKING?

Detail banking should begin by establishing guidelines, policies, and procedures for implementing a system. A process for review of past documents and files should be established with guidelines for evaluating existing and new details. Success of this system will depend on overall program management and the policies governing use of standard details. To achieve success requires the assignment of one individual to carry out responsibilities associated with detail banking.

Most offices have a number of potential details that can serve as basis for establishing a detail banking system. Three important steps can lead the design professional into this system:

1. Establish selection criteria to rank the order in which a detail category is selected for filing.

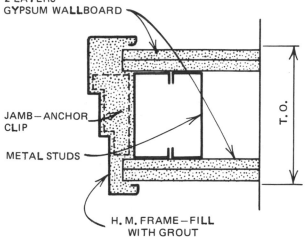

Figure 2-3 Door Jamb Detail (Smith, Hinchman & Grylls Associates, Inc., Detroit, Michigan).

2. Select categories of details that have the greatest potential for reuse, such as door details; partition details; a series designed for 1, 2, 3, and 4 hour fire rated walls; cabinet details; window details; stair details. Several standard details are shown in Figures 2-2, 2-3, 2-4, 2-5, 2-6, and 2-7 to demonstrate the example categories.
3. Develop a standard format for drawing, filing, and retrieving details.

2-3.1 Multidiscipline design considerations for detail banking.

The preceding example categories have been selected from the architectural discipline. However, similar examples of reusable detail categories can be taken from related disciplines in the construction industry. For example, engineering drawings for mechanical and electrical systems can also provide many standard details. Landscape architects, interior designers, and industrial designers should also review their historic files to select reusable or standard details for banking. The procedures, guidelines, and system concepts presented will be applicable to other disciplines using drawings and details to communicate construction information.

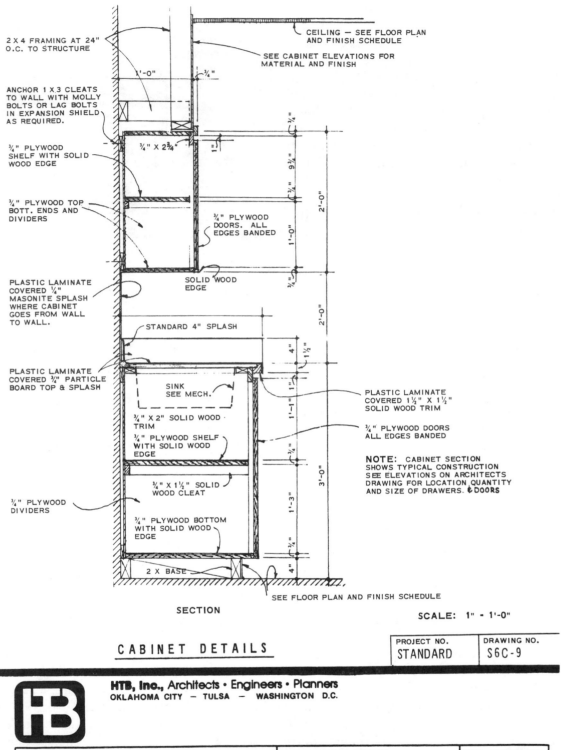

2 X 4 FRAMING AT 24" O.C. TO STRUCTURE

CEILING — SEE FLOOR PLAN AND FINISH SCHEDULE

SEE CABINET ELEVATIONS FOR MATERIAL AND FINISH

ANCHOR 1 X 3 CLEATS TO WALL WITH MOLLY BOLTS OR LAG BOLTS IN EXPANSION SHIELD AS REQUIRED.

¾" PLYWOOD SHELF WITH SOLID WOOD EDGE

¾" X 2¾"

¾" PLYWOOD TOP BOTT. ENDS AND DIVIDERS

¾" PLYWOOD DOORS. ALL EDGES BANDED

PLASTIC LAMINATE COVERED ¼" MASONITE SPLASH WHERE CABINET GOES FROM WALL TO WALL.

SOLID WOOD EDGE

STANDARD 4" SPLASH

PLASTIC LAMINATE COVERED ¾" PARTICLE BOARD TOP & SPLASH

SINK SEE MECH.

PLASTIC LAMINATE COVERED 1½" X 1½" SOLID WOOD TRIM

¾" X 2" SOLID WOOD TRIM

¾" PLYWOOD SHELF WITH SOLID WOOD EDGE

¾" PLYWOOD DOORS ALL EDGES BANDED

¾" X 1½" SOLID WOOD CLEAT

NOTE: CABINET SECTION SHOWS TYPICAL CONSTRUCTION SEE ELEVATIONS ON ARCHITECTS DRAWING FOR LOCATION QUANTITY AND SIZE OF DRAWERS & DOORS

¾" PLYWOOD DIVIDERS

¾" PLYWOOD BOTTOM WITH SOLID WOOD EDGE

2 X BASE

SEE FLOOR PLAN AND FINISH SCHEDULE

SECTION

SCALE: 1" = 1'-0"

CABINET DETAILS

PROJECT NO.	DRAWING NO.
STANDARD	S6C-9

HTB, Inc., Architects • Engineers • Planners
OKLAHOMA CITY — TULSA — WASHINGTON D.C.

REMARKS:				DESCRIPTIVE TITLE		STD. SHEET NO.
MARK	REVISIONS	DATE	BY	DRAWN		
				CHECKED		
				DESIGNER		
				JOB CAPT.		
				PROJ. MGR.		
				DIRECTOR		
				ASST. DIRECTOR		

Figure 2-4 Cabinet Details (HTB, Inc., Architects/Engineers/Planners, Oklahoma City, Oklahoma).

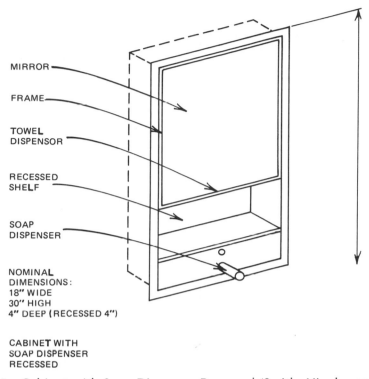

MIRROR

FRAME

TOWEL
DISPENSOR

RECESSED
SHELF

SOAP
DISPENSER

NOMINAL
DIMENSIONS:
18" WIDE
30" HIGH
4" DEEP (RECESSED 4")

CABINET WITH
SOAP DISPENSER
RECESSED

Figure 2-5 Cabinet with Soap Dispenser Recessed (Smith, Hinchman & Grylls Associates, Inc., Detroit, Michigan).

2-4 DEVELOPING A DETAIL GRADING SHEET

A detail grading sheet should be constructed prior to selecting and storing working drawing details. Each detail should be thoroughly reviewed and screened for potential errors as well as for constructibility. The grading process should seek input from construction and manufacturing representatives to help ensure development of quality details.

Grading issue should represent critical material performance requirements and construction methods. New materials should be evaluated and approved before utilized in design of a construction detail. Specific items should incorporate learning experiences from common errors presently observed in building failures. The Material Evaluation Form and Grading Sheet shown in Figures 2-8 and 2-9 represent the suggested format to be constructed by the firm contemplating development of a *detail data bank*.

2-4.1 Detail review, evaluation, and selection for the banking system.

The value of a standardized detail banking program is directly related to the review and evaluation process. Therefore some form of grading sheet and historic data card becomes critical for recording the status of a specific detail. Reuse of a detail should only be considered after a thorough evaluation. A grading sheet can also serve as an excellent checklist in a quality control check of the working drawings.

2-5 DETAIL SELECTION PROCEDURES

The best method of organizing a *detail data bank* is to study the past history of the design firm. First begin by going back through working drawing details generated for past project types. Review all projects that might be filed for a particular type

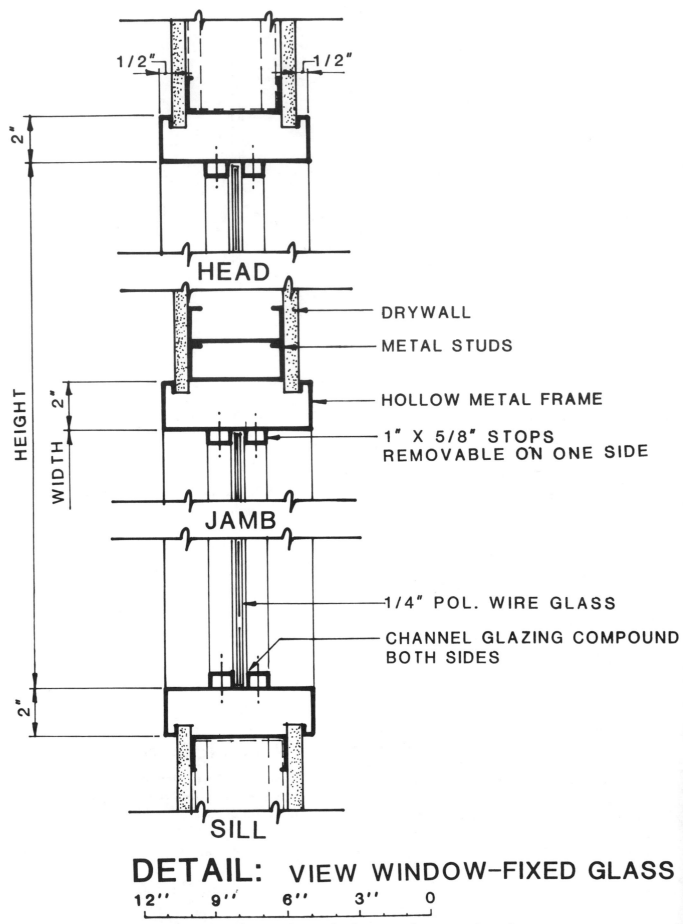

Figure 2-6 View Window with Fixed Glass (Gresham, Smith and Partners, Nashville, Tennessee).

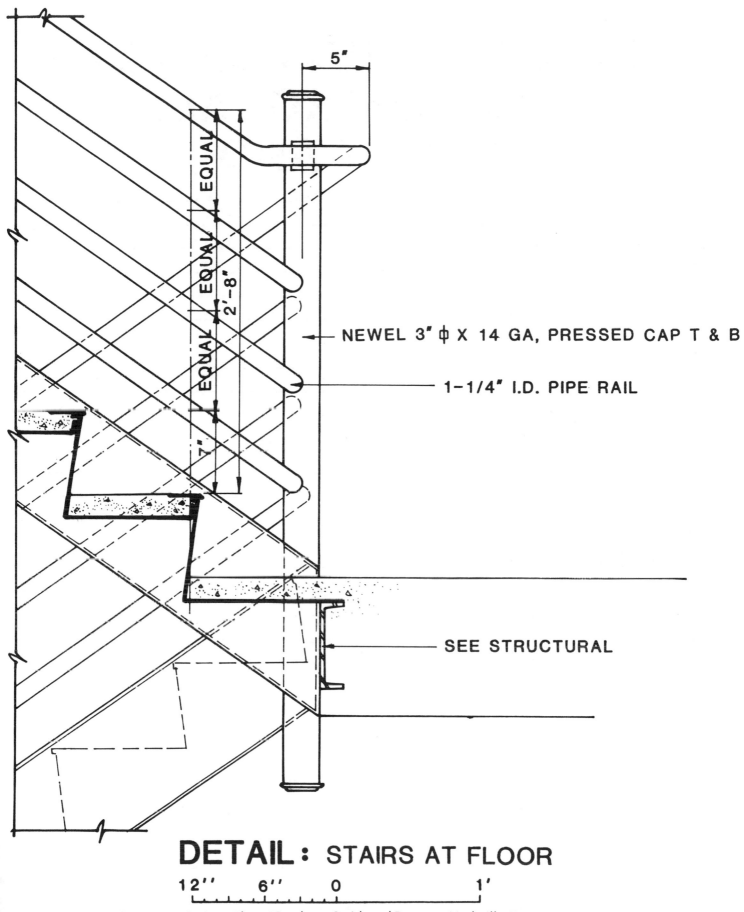

5"

EQUAL
EQUAL
EQUAL
2'-8"
EQUAL
EQUAL
7"

NEWEL 3" φ X 14 GA, PRESSED CAP T & B

1-1/4" I.D. PIPE RAIL

SEE STRUCTURAL

DETAIL: STAIRS AT FLOOR

12'' 6'' 0 1'

Figure 2-7 Stairs at Floor (Gresham, Smith and Partners, Nashville, Tennessee).

MATERIAL EVALUATION AND APPROVAL DATA CARD

Approval No. _____

Material Description _____ [ ]

Manufacturer _____

Manufacturer's Certification Statement _____

Data Received _____

Test Information Available _____

Test Required _____

_____

_____

Response to Test

_____ Excellent _____ Good _____ Poor

Evaluation Statement _____

_____

_____

Approval for _____

Date Approved _____

Special Conditions _____

Approval Period _____

Approved by _____

Field Evaluation Data _____

Problems _____

_____

Recommend Further Testing _____

_____

Retrieval Descriptors

Figure 2-8 Material Evaluation and Approval Data Card.

DETAIL GRADING SHEET

Detail Description _____ No. _____

Evaluation Issues	5 Excellent	4 Good	3 Average	2 Fair	1 Poor
Material Compatibility					
Aesthetics—Overall Design					
Corrosion Control					
Expansion Control					
Contraction Control					
Weathering Capability					
Constructibility					
Structural Capabilities					
Material Availability					
Thermal Characteristics					
Water Permeability					
Acoustical Qualities					
Fire Safety					
Maintainability					
Tolerances					
Total Points					

Detail Evaluation

☐ Acceptable

☐ Unacceptable

☐ Needs Refinement

☐ Requires Further Evaluation and Testing

Comments _____

_____

_____

_____

_____

For further expansion of the evaluation issues, review systematic evaluation process
identified in document on Construction Material Evaluation and Selection—
A Systematic approach by Harold Rosen and Philip M. Bennett. John Wiley & Sons, C. 1979

Figure 2-9 Detail Grading Sheet.

of facility. Once drawings are separated, review specific details. Details that have a high reoccurrence should be separated out and categorized for future review and study. Once all details have been reviewed, then project the possible categories required to file each detail. Details with the greatest reuse potential will be quickly spotted for detail banking. Based on the firm's projected growth, it will be important to select categories that represent past and present development of a particular detail. Code each detail with a numbering system that indicates the frequency this detail was reused over a given period of time.

For example, a frequency reuse graph or code can be designated on each detail selected from the historic files. This designation can be shown with a number or a graphically drawn scale to represent detail reuse. The frequency graph approach shown in Figure 2-10 provides the flexibility to approximate the number of times a detail was reused over a given time period. A low, medium, and high frequency scale creates the categories for dividing historic details into levels of reuse. The detail category with greatest reuse potential should be given the highest priority for input to the detail banking system.

Establishing a frequency scale is necessary to represent levels of detail reuse. This scale should be based on firm size, project types, and recurrence of similar projects. The shaded area in the graph represents the approximate frequency a detail was reused. This area can be easily altered if detail reuse changes. (A frequency graph can also be placed on the Detail Data Card shown in Figure 3-26.) It is recommended that actual project data such as project name, title, and detail location be recorded on the card each time a detail is used on a new project. This information will serve as a permanent record for detail research on future projects.

Different methods of indexing details into reuse categories can be devised by an innovative indexer. In some firms color coding is used to represent different levels of detail reuse. The following example, presented in *The Paper Plane*, June 1979, highlights a method used by an Arizona firm [2].

Color-Coding Frequency

Detail information is recorded on a card that indicates frequency of reuse by color coding. Cards for all details are filed for future reference. For example:

Low = 0–5 times per year
Medium = 5–25 times per year
High = 25–100 times per year

Frequency Graph

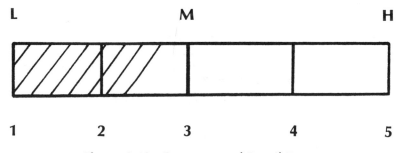

Figure 2-10 Frequency of Detail Reuse.

1. Repetitive type = green color code.
2. Reference type = yellow color code.

Graphic forms of representation can work similar to bar chart records used to show percent of drawing completion. Critical issues for any method are (1) the ease with which it can be used, (2) how simple it is to understand, and (3) whether it provides desired results for the user.

Many times design professionals don't realize the number of times details are reused on similar projects. From an actual survey of historic drawing files, one can pinpoint repetition in detail development. To save time and money in drawing production, it is necessary to minimize repetition by standardizing details that are similar in function, performance, and aesthetics. Some of the typical standard detail categories in construction drawings are shown in the following listing.

Types of Standard Details

Exterior	Interior
Foundations	Cabinets
Walls	Stairs
Roofs	Partitions
Windows	Ceilings
Doors	Doors
Plazas	Railings
Decks	Chimneys
Walks	Fireplaces
Drives	Trim
Planters	Expansion and
Fountains	control joints
Signs	Tack and chalk
Lighting	boards
Railings	
Expansion and	
control joints	

Other design-oriented disciplines should analyze their production drawings to determine the detail categories most appropriate for banking. To aid in development of a standard detail collection, the design professional should consider outside sources of information. Potential sources to obtain developed details are *manufac-*

turers, professional associations, photographs, and *architectural and engineering consultants.*

2-6 MANUFACTURERS' STANDARD DETAILS

Descriptive literature on standard products being specified by many firms contains the manufacturer's details. For many products these details need little or no modifications to be acceptable for construction drawings. Most manufacturers are pleased to have their details reused by the design professional. The examples shown in Figures 2-11 and 2-12 demonstrate standard door frame information made available by many manufacturers.

2-7 PHOTOGRAPHIC DETAILS

The power of the camera has been underestimated when it comes to production of working drawings. Photographic details can be acquired by using photographs of *products and equipment, mock-ups and assemblies, or actual construction events.* Some of the major benefits of a photographic detail are:

1. Reduced production time.
2. Overall cost savings.
3. Realism in construction communication.
4. Limited training required.

The selected photographic details shown in Figures 2-13 and 2-14 represent the wide variety of opportunities for obtaining standard details.

2-8 ASSOCIATION DETAILS

Many industry organizations and associations fund research and development programs that result in recommended standards and details. These details represent composite knowledge and

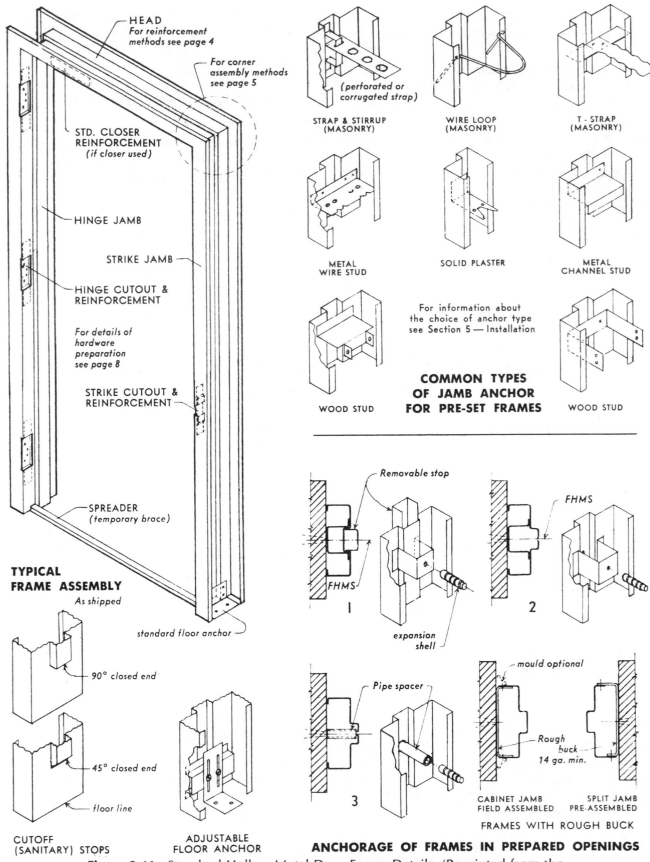

HEAD
For reinforcement methods see page 4

For corner assembly methods see page 5

STD. CLOSER REINFORCEMENT
(if closer used)

HINGE JAMB

STRIKE JAMB

HINGE CUTOUT & REINFORCEMENT

For details of hardware preparation see page 8

STRIKE CUTOUT & REINFORCEMENT

SPREADER
(temporary brace)

TYPICAL FRAME ASSEMBLY
As shipped

standard floor anchor

90° closed end

45° closed end

floor line

CUTOFF (SANITARY) STOPS

ADJUSTABLE FLOOR ANCHOR

STRAP & STIRRUP (MASONRY)
(perforated or corrugated strap)

WIRE LOOP (MASONRY)

T - STRAP (MASONRY)

METAL WIRE STUD

SOLID PLASTER

METAL CHANNEL STUD

WOOD STUD

For information about the choice of anchor type see Section 5 — Installation

COMMON TYPES OF JAMB ANCHOR FOR PRE-SET FRAMES

WOOD STUD

Removable stop

FHMS

FHMS

expansion shell

1

2

Pipe spacer

3

mould optional

Rough buck 14 ga. min.

CABINET JAMB FIELD ASSEMBLED

SPLIT JAMB PRE-ASSEMBLED

FRAMES WITH ROUGH BUCK

ANCHORAGE OF FRAMES IN PREPARED OPENINGS

Figure 2-11 Standard Hollow Metal Door Frame Details. (Reprinted from the *Hollow Metal Technical and Design Manual*, Copyright 1977, with permission of the Hollow Metal Manufacturers Association, a Division of the National Association of Architectural Metal Manufacturers, Chicago, Illinois.)

FRAME DETAILS

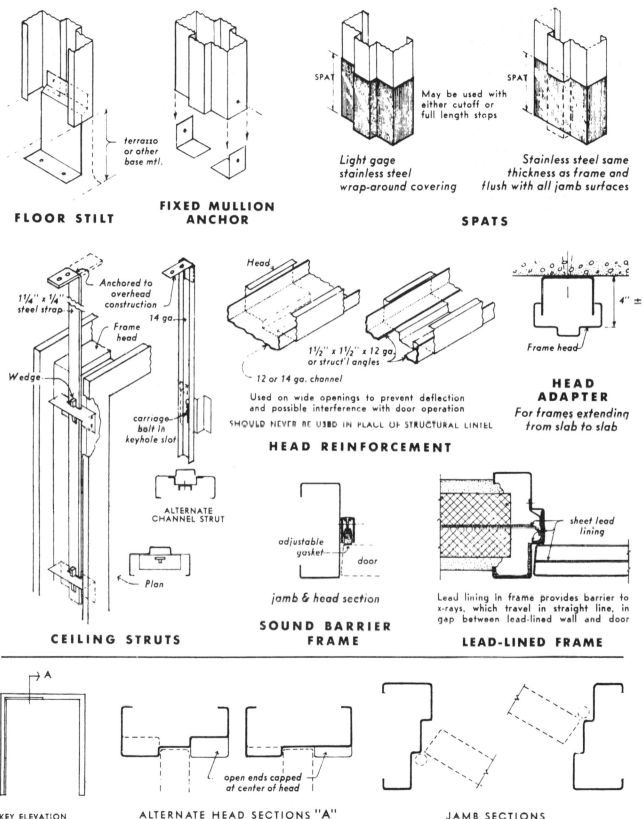

FLOOR STILT

FIXED MULLION ANCHOR

SPAT — May be used with either cutoff or full length stops

Light gage stainless steel wrap-around covering

SPAT

Stainless steel same thickness as frame and flush with all jamb surfaces

SPATS

Anchored to overhead construction

1¼" x ¼" steel strap

14 ga.

Frame head

Wedge

carriage bolt in keyhole slot

ALTERNATE CHANNEL STRUT

Plan

CEILING STRUTS

Head

1½" x 1½" x 12 ga. or struct'l angles

12 or 14 ga. channel

Used on wide openings to prevent deflection and possible interference with door operation

SHOULD NEVER BE USED IN PLACE OF STRUCTURAL LINTEL

HEAD REINFORCEMENT

adjustable gasket

door

jamb & head section

SOUND BARRIER FRAME

Frame head

4" ±

HEAD ADAPTER

For frames extending from slab to slab

sheet lead lining

Lead lining in frame provides barrier to x-rays, which travel in straight line, in gap between lead-lined wall and door

LEAD-LINED FRAME

A

KEY ELEVATION

open ends capped at center of head

ALTERNATE HEAD SECTIONS "A"

JAMB SECTIONS

DETAILS OF DOUBLE EGRESS FRAME

Figure 2-12 Standard Hollow Metal Door Frame Details. (Reprinted from the *Hollow Metal Technical and Design Manual*, Copyright 1977, with permission of the Hollow Metal Manufacturers Association, a Division of the National Association of Architectural Metal Manufacturers, Chicago, Illinois.)

3'-0"

Figure 2-13   Recessed Drinking Fountain.

experience of many professionals active in various associations. Design professionals can save time and money by utilizing existing details, research, and established standards made available by associations in the form of manuals. The detail sheets shown in Figures 2-15 and 2-16 have been selected from two active associations to represent the type and format typically used to present design information to the construction industry [3], [4].

2-9 ARCHITECTURAL AND ENGINEERING DETAILS

Standardization of details has been happening in some firms for a number of years. Until recently, however, benefits have not been widespread or fully realized by many in the design profession. Production time and costs are now causing designers to look at repetitive drawings and make every effort to minimize labor-intensive activities. This is motivating many organizations to create standard reusable details that have built-in

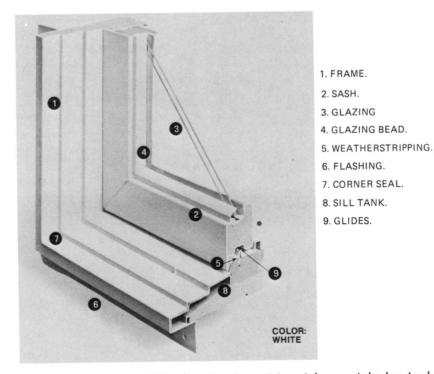

1. FRAME.

2. SASH.

3. GLAZING

4. GLAZING BEAD.

5. WEATHERSTRIPPING.

6. FLASHING.

7. CORNER SEAL.

8. SILL TANK.

9. GLIDES.

COLOR: WHITE

Figure 2-14   Vinyl Coated Window Section. (Material copyright by Andersen Corporation, Bayport, Minnesota.)

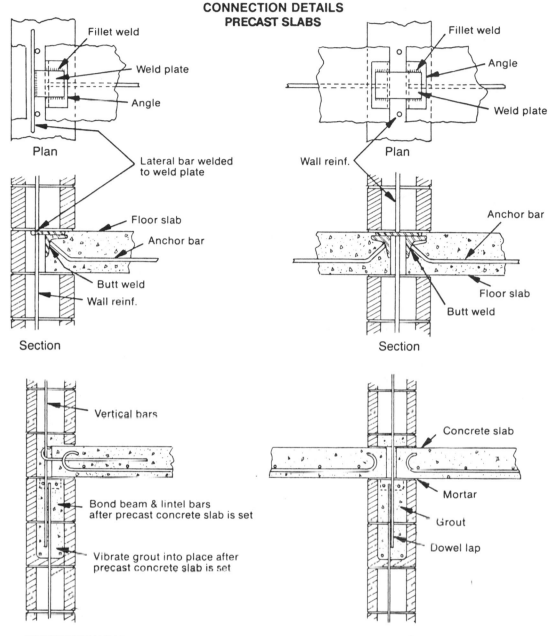

CONNECTION DETAILS PRECAST SLABS

Plan — Fillet weld, Weld plate, Angle

Lateral bar welded to weld plate

Section — Floor slab, Anchor bar, Butt weld, Wall reinf.

Plan — Fillet weld, Angle, Weld plate

Wall reinf.

Section — Anchor bar, Floor slab, Butt weld

Vertical bars

Bond beam & lintel bars after precast concrete slab is set

Vibrate grout into place after precast concrete slab is set

Concrete slab

Mortar

Grout

Dowel lap

EXTERIOR WALL

INTERIOR WALL

Figure 2-15 Connection Details for Precast Slabs (National Concrete Masonry Association, Herndon, Virginia).

quality control and general application in construction. As this program grows, more and more details will become available to the design professional. Some firms are already making plans to exchange or provide for the purchase of standard details. When this opportunity arrives, the design professional needs to consider carefully the quality and format of each available detail. Details shown in Figures 2-17 and 2-18 represent the type of standardization in linework, format, and notation needed to create consistency in a future detail exchange program.

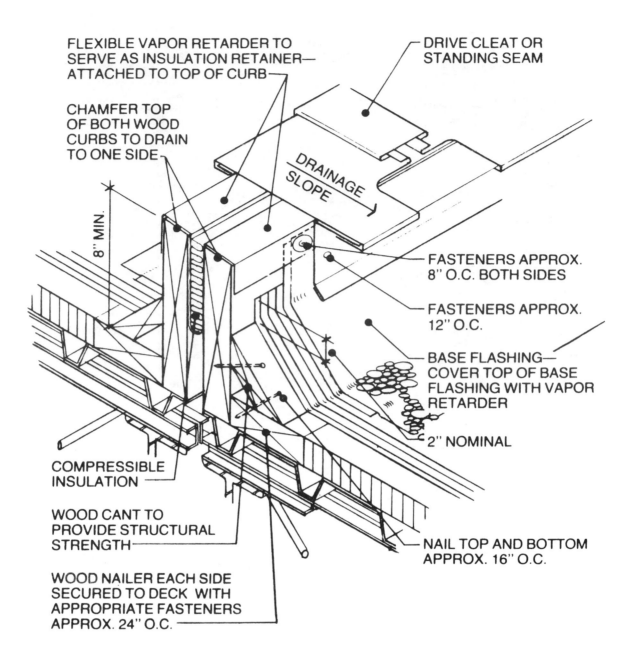

FLEXIBLE VAPOR RETARDER TO
SERVE AS INSULATION RETAINER—
ATTACHED TO TOP OF CURB

DRIVE CLEAT OR
STANDING SEAM

CHAMFER TOP
OF BOTH WOOD
CURBS TO DRAIN
TO ONE SIDE

DRAINAGE SLOPE

8" MIN.

FASTENERS APPROX.
8" O.C. BOTH SIDES

FASTENERS APPROX.
12" O.C.

BASE FLASHING—
COVER TOP OF BASE
FLASHING WITH VAPOR
RETARDER

2" NOMINAL

COMPRESSIBLE
INSULATION

WOOD CANT TO
PROVIDE STRUCTURAL
STRENGTH

NAIL TOP AND BOTTOM
APPROX. 16" O.C.

WOOD NAILER EACH SIDE
SECURED TO DECK WITH
APPROPRIATE FASTENERS
APPROX. 24" O.C.

1980

C-1

NOTE:

THIS DETAIL ALLOWS FOR BUILDING MOVEMENT IN BOTH DIRECTIONS.
IT HAS PROVEN SUCCESSFUL WITH MANY CONTRACTORS FOR MANY
YEARS.

Figure 2-16 Roof Expansion Joint (National Roofing Contractors Association,
Chicago, Illinois).

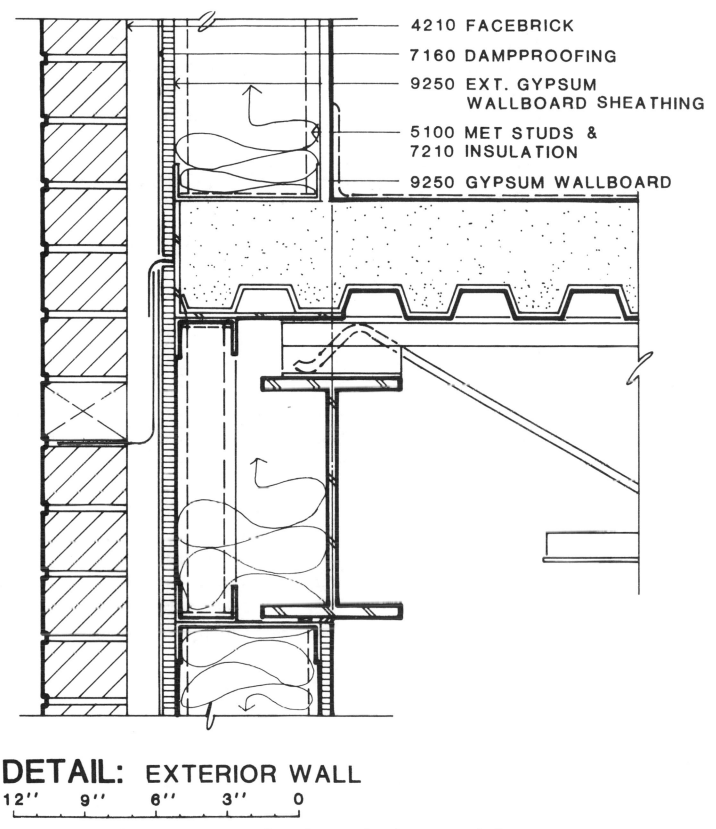

4210 FACEBRICK

7160 DAMPPROOFING

9250 EXT. GYPSUM WALLBOARD SHEATHING

5100 MET STUDS & 7210 INSULATION

9250 GYPSUM WALLBOARD

DETAIL: EXTERIOR WALL

12''   9''   6''   3''   0

Figure 2-17   Exterior Wall (Gresham, Smith and Partners, Nashville, Tennessee).

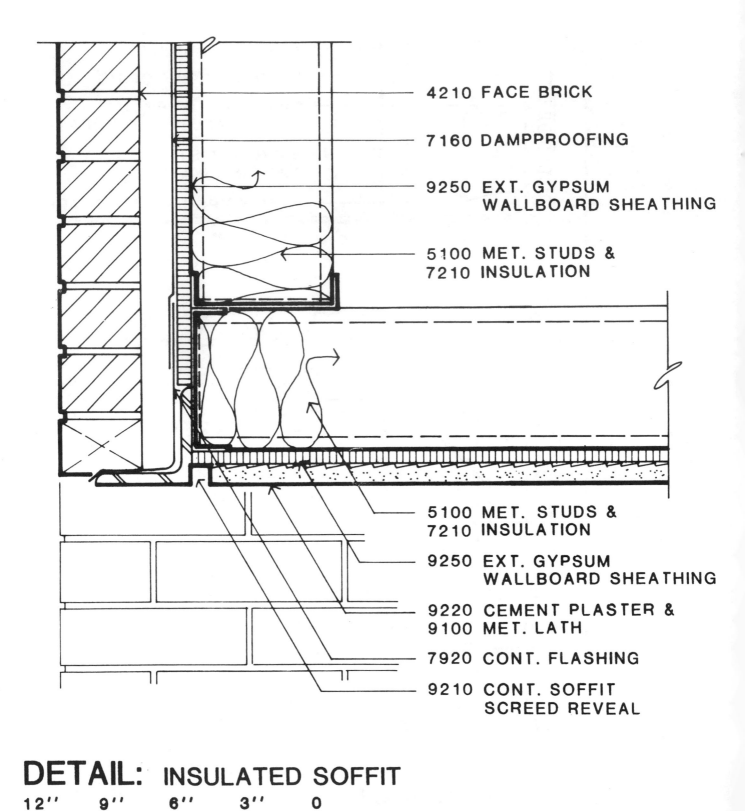

4210 FACE BRICK

7160 DAMPPROOFING

9250 EXT. GYPSUM
WALLBOARD SHEATHING

5100 MET. STUDS &
7210 INSULATION

5100 MET. STUDS &
7210 INSULATION

9250 EXT. GYPSUM
WALLBOARD SHEATHING

9220 CEMENT PLASTER &
9100 MET. LATH

7920 CONT. FLASHING

9210 CONT. SOFFIT
SCREED REVEAL

DETAIL: INSULATED SOFFIT

12''   9''   6''   3''   0

Figure 2-18 Insulated Soffit (Gresham, Smith and Partners, Nashville, Tennessee).

22

THREE
PREPARATION OF STANDARD DETAILS

3-1 IMPORTANCE OF QUALITY CONTROL

The preparation process for today's construction details varies greatly between projects and offices. Many details are quickly produced by using a variety of scales, techniques, and abstract symbols to convey graphic information. Lack of control over mass production of details for many different projects has reduced personal recognition for a solution to a unique design problem. As construction details become more impersonal, there is less concern for maintaining historic records and files as seen in earlier periods of architecture.

Many of today's drawings are mass produced, filed, and forgotten. Legal problems arise all too often because details have been drawn without proper research and material investigation to arrive at a good design decision. Hasty decision making, minimal field feedback, and reduced personal input have set the stage for many legal problems. A study of construction claims published in *Guidelines for Improving Practice* (Victor O. Schinnerer and Company, 1971), indicates that many errors in graphic communication are the result of three shortcomings [5]:

1. Inadequate or poor communication among the designer, the draftsman and the specifications writer.
2. Lack of "in the field/on the job" experience by the draftsman. Often, the draftman has never had an opportunity to observe the performance of actual construction procedures. This lack of experience often prevents him from fully understanding the manner in which the design is to materialize.

3. Superficial review, or worse, none at all of the working drawings by a principal or qualified supervisor.

□

3-2 PLANNING FOR DETAIL PRODUCTION

3-2.1 Select essential information.

Detail design that reflects good construction practices is a must in detail banking. All information utilized in production of details must be evaluated in accordance with present construction technology. Structuring an office library will aid in creating a system to supply information for detail development. Current construction information combined with user requirements is essential to achieving effective details. Every effort should be put forth to use high-quality, current information that promotes clear, precise working drawing details that can be easily understood during construction.

3-2.2 Use language consistent with working drawings and specifications.

Effective communication should be based on a construction language that is generated by all member disciplines of the design profession. Terminology considered standard in the field should be documented and identified in the building construction thesaurus which can be referred to as details and working drawings are prepared. The language selected for each detail should also

RECOMMENDED USE	PRESENT VARIATIONS
Plastic Laminate	Formica, Laminated Plastic, and Textolite
Gypsum Wallboard	Sheetrock, Drywall Plasterboard, and Gypsum Board
Chalkboard	Slate and Blackboard

Figure 3-1 Standardizing Terminology.

be coordinated with the specifications department so that inconsistencies in use will not occur. Cross-checking will be important toward achieving a consistent and uniform set of working drawings and specifications. Only generic terms should be used in standard detail systems. Terminology that represents trade names or specific industry terms that may change frequently should be avoided. The example in Figure 3-1 gives terms recommended for use in a construction language.

understanding of present construction technology so that details can be designed and built utilizing construction practices that attain the highest quality of building performance. Building failures can result when past experience is not utilized in design and construction. For example, expanding brick can cause a failure in an exterior wall if the proper shelf angles and expansion joints are not provided. The detail in Figure 3-2 demonstrates the potential for relief at horizontal expansion joints.

3-2.3 Develop a balance between abstract and realistic drawing so users can understand the information with least amount of time and effort.

Graphic representation should not be so abstract that users of construction documents cannot understand the intent of the information. Field tests should be conducted to determine the proper balance between realism and abstraction in detail development. Guidelines for achieving proper graphic balance in detail development should be identified for consistency.

3-2.5 Be precise as to scale, size of components, and relationship to adjacent construction conditions.

In grading each detail, review all components for scale, dimensions, and overall graphic communication. Critical material relationships should be noted for possible inconsistencies with construction technology. Aesthetic consideration should also be given to material uses and relationships. For example, many problems can be identified where corrosion between materials develops as a result of improper detailing.

3-2.4 Detail material relationships compatible and obtainable with present technology.

Building material evaluation and selection is critical in developing successful standard details. Developing an understanding of material relationships is also necessary to achieve workable details in the field. It is important to gain an

3-2.6 Develop a design approach that is consistent and integrated with the entire structure so that each detail contributes to the total system.

Design firms should review and agree on an overall design approach to aesthetic considerations in building design. All members of the

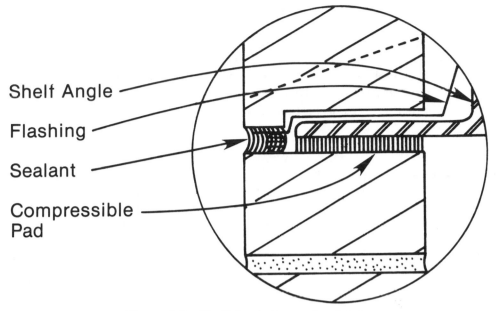

Shelf Angle
Flashing
Sealant
Compressible
Pad

Figure 3-2 Shelf Angle Expansion Joint.

design team should coordinate design approaches in order to arrive at a consistent and pleasing architectural relationship between all parts of a building. Since architectural designers come from different training backgrounds, it is important to coordinate their design concepts so as to achieve an integrated approach in the final design of each structure. An example would be coordinating the design concepts of several members whose design background consists of training under several well-known architects with recognized design philosophies. An approach that allows designers to work independently, and is conducive to individual creativity will tend to fragment the entire building design process and create problems in structuring a unified detail banking system. It is recommended that new members of the firm receive special in-house training on coordination responsibilities for achieving a total integrated building design.

It is important to remember that good clear communication and coordination will save time, money, and negative feedback during the construction process. Time and effort spent on planning and coordinating early stages of detail and working drawing development will be paid back by the time of building completion. Many construction failures can be attributed to the lack

of preparation given to graphic communication. Major construction claims have resulted from inaccurate graphic information being communicated to field personnel.

3-3 THE DETAIL DEVELOPMENT PROCESS

Effective construction details can be attained when individual and departmental strengths are developed in two areas: (1) in construction technology through education and exposure to construction procedures and (2) in departmental planning and organization to establish a systematic design process.

The first area of concentration involves both an individual and firm responsibility to develop educational opportunities that expand knowledge of construction technology. High-quality working drawings result from good planning, research, and organization of information in a logical manner that is acceptable to the construction industry. An effective program must reflect the level of communication, skills, and technology achieved during the time frame of the design process.

The second area of concentration requires a

departmental effort to create a design process that is structured and systematic. An analysis of a firm's work, goals, and objectives will help to crystalize a comprehensive program for detail development. Organizational activities in five major task areas are necessary to bring about an effective detail development process as presented in Figure 3-3 and 3-4.

3-4 STANDARD DETAIL REQUIREMENTS

To develop effective construction details requires care in preparation followed by thorough evaluation. The quality of standard or reusable details is heavily dependent on the designer's depth of technical experience in construction materials. Criteria for achieving success in the detail devel-

Need
- **Purpose**
- **Functions**
- **Requirements**

Research
- **Material Technology**
- **Material Relationships**
- **Construction Techniques**
- **Field Feedback**
- **Past Problems**
- **Expertise**

Review-Coordination-Approval
- **Architectural**
- **Mechanical**
- **Electrical**
- **Structural**
- **Consultants**
- **Specifications**
- **Manufacturers**
- **Contractors**

Standards
- **Scale**
- **Materials**
- **Linework**
- **Lettering & Dimensioning**
- **Freehand/Mechanical**
- **Abbreviations**
- **Terminology**
- **Numbering**
- **Ink or Pencil**

Development
- **Aesthetics**
- **Compatibility of Materials**
- **Building Codes**
- **Climate**
- **Fabrication**
- **Tolerances**
- **Installations**
- **Material Availability**
- **Costs**

Figure 3-3 Detail Development Process.

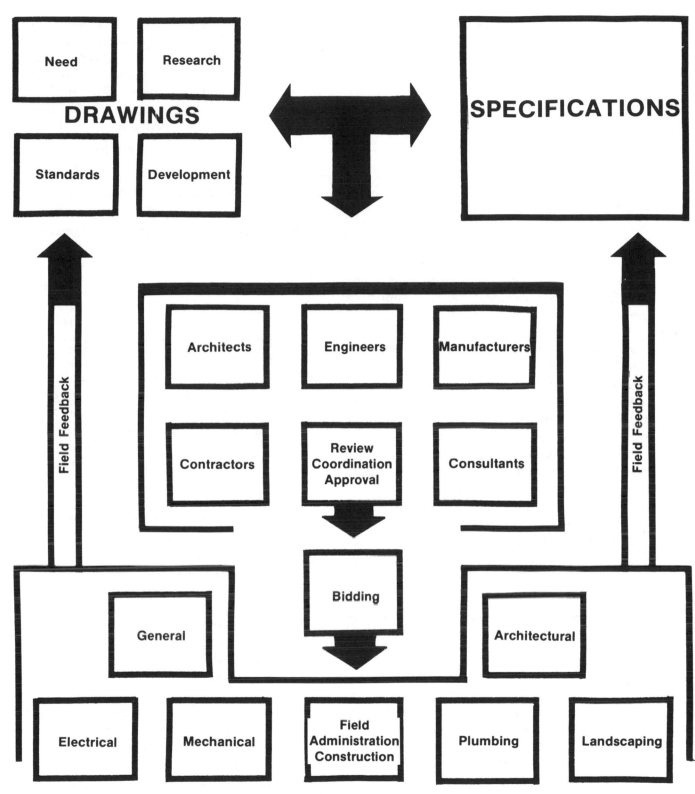

Figure 3-4 Detail Development.

opment process can be measured with the following attributes:

▷ Functional
▷ Buildable
▷ Technically sound
▷ Current
▷ Clarity

To fulfill these attributes, a searching out of standardized activities should be considered.

3-4.1 Identify critical areas of graphic communication that require standardization in your office.

Standardized details require that a consistent and standard method of drafting be used as they are developed. Communication methods must be identified so that all members of the design team are working with consistent and uniform techniques for documenting information on standard details. An analysis of present methods should be undertaken before establishing new guidelines. A procedure and policy manual should then be prepared within the office before proceeding to develop standard details. The examples that follow and the guidelines shown in Figures 3-5 and 3-6 were prepared to control the detail development process.

Guidelines on Producing Drawings For A Project*

1. What to draw where.

    a. On small-size drawings include plan details; wall section details; major component systems; doors, frames, stairs, millwork, etc.; components drawing door types, cabinets; small parts; schedules.

*Source: Thorsen & Thorshov Associates Inc., Architects/ Planners, Minneapolis, Minnesota.

b. On large-size drawings include floor plans; elevations (interior and exterior); building sections; reflected ceiling plans (same scale as floor plans); wall sections (minimal details); area "blow ups"; details that by their nature or information value will not function on small sheets.

2. If you can't justify creating a new drawing, *don't draw!*

    a. Scan SSD index for related drawing(s) that can be used.

    b. Retrieve mylar from SSD file drawer, make photocopy, and return mylar to SSD file drawer.

    c. Use photocopy as "original" for project. It can be photocopied and assembled into book of SSDs for the project or photocopied as a sticky back on large drawings.

    d. On photocopy "original" tie in vertical dimensions of critical planes such as floors, top of steel, exterior grade. Show plan grids and tie to horizontal dimensions. Avoid a lot of duplication of dimensional information (obviously, some duplication is necessary for clarity).

    e. On photocopy "original" add reference numbers to "bubbles." Circles at bottom of sheets are for indicating related drawings other than those traditionally shown on the detail. The right-hand circle "bubble" is used for "backwards" referencing to a drawing on which the detail occurs. These references provide additional information not possible by traditional methods. The lower right square is reserved for the number that identifies the detail for the project. Numbering of details for a project is controlled by whatever systems the project architect wishes to use.

    f. In the event a drawing needs to be modified, the modifications should be red-lined onto the old file photocopy in the three-ring binder. After modifica-

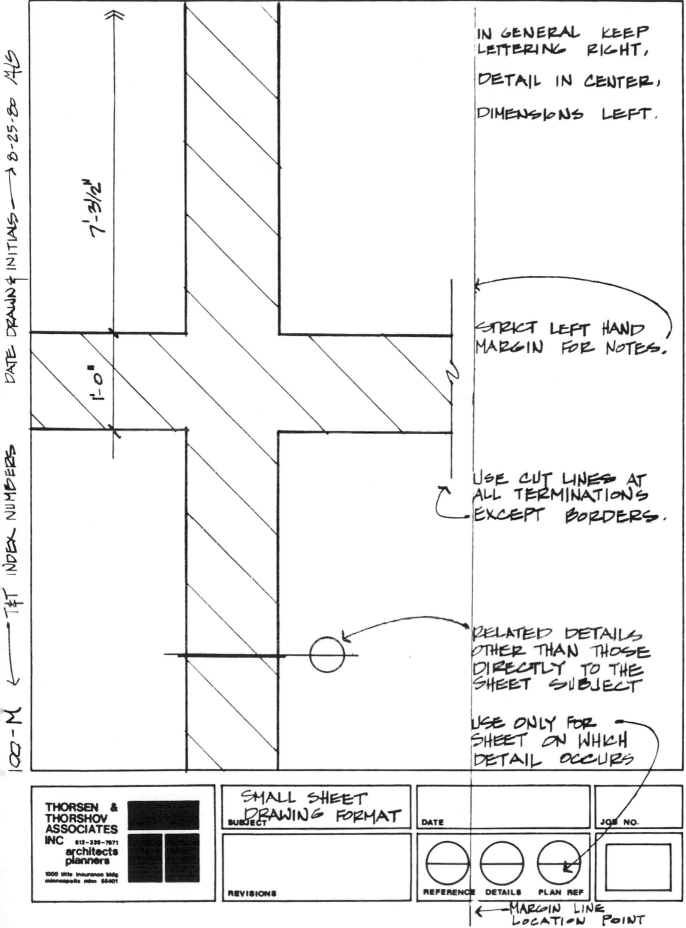

Figure 3-5 Small Sheet Drawing Format (Thorsen & Thorshov Associates, Inc., Minneapolis, Minnesota).

tions are made to the mylar, insert a photocopy in the three-ring binder and return the mylar to the file drawer. The retained red-lined copies will show the stepwise improvements of details over time and will provide valuable information in evaluating future detailing.

3. How to create a new drawing.

   a. Obtain preprinted 8½ × 11 inch mylar sheets.

   b. For format; refer to SSD 100M. Consistency in organization is mandatory in producing contract documents for a job; as more jobs are produced, there will be sets of sheets from many jobs. The details produced on one job will be compatible with others if the format and following elements are consistent:

      ○ *Layout* Complete details to the lines indicated; don't arbitrarily stop lines. Stop all details with cut lines.

         Notes Keep lettering within right margin and dimensions to the left.
      ○ Use extended letters of constant height (compressed letters are harder to read, especially on reduced prints). Avoid multiple heights and slanted styles.
      ○ *Scale* Draw details at a scale applicable to the entire series of details (scale to be determined by experience and by existing details).

   c. Leave reference "bubbles" and lower right square blank.

   d. Select a meaningful sheet title. Titles are important as a source of information in the future. Titles should be specific yet broad enough to be usable on other projects.

   e. Determine permanent index filing number (from step 2e) and write number in left margin.

   f. Make two photocopies; then file mylar in SSD file drawer.

   g. Insert one photocopy into three-ring binder, and add title of drawing to index.

   h. Treat second photocopy as "original" for project and proceed with steps 2c, 2d, and 2e.

   □

3-4.2 Determine the system to be used in placing details on final drawings.

New techniques and methods are now available for preparing standard details for working drawing sheets. It is important to determine methods to be used in applying details to working drawing sheets before selecting a procedure for preparing standard details. In some systems a variety of techniques will be used to apply details to working drawings. This will require proper preparation of each original so that reproducibles can be obtained for making uniform copies. The techniques identified in Figure 3-7 have proved successful for many firms.

3-4.3 Standardize graphic representation in production of working drawings.

Graphic communication must be standardized using all acceptable and proven techniques for achieving successful reproduction of construction information. To develop this level of standardization, it will require extensive researching of the current available techniques. A state-of-the-art study should be conducted before developing your office procedures manual. The examples shown in Figures 3-8, 3-9, 3-10, 3-11, 3-12, and 3-13 demonstrate organized standards and identify sources of valuable data for selecting appropriate standardization techniques in graphic communication. Critical areas of standardization are:

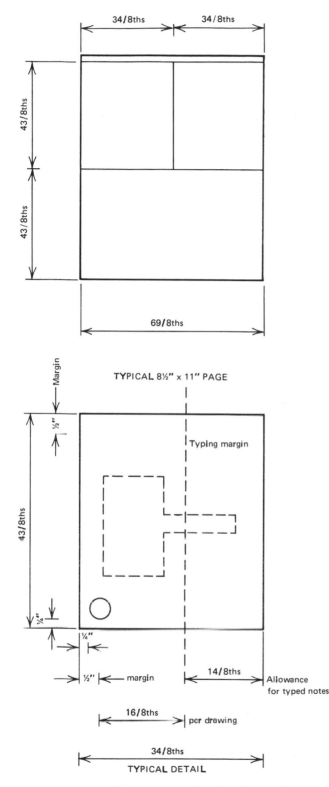

Standardized Detail Drawing Format

Details are to have the following final dimensions:

▷ Single detail—34/8ths × 43/8ths (4¼″ × 5⅜″).

▷ Double horizontal detail—68/8ths by 43/8ths (8½″ × 5⅜″).

▷ Double vertical detail—34/8ths × 88/8ths (4¼″ × 11″).

▷ Full sheet—68/8ths × 88/8ths (8½″ × 11″).

Draw detail at required size on 1000H (clearprint with fade-out grid), with margins and circles as shown. An 8½″ × 11″ sheet should be completely filled for most economical use. *The corners of each detail should be clearly marked.*

Position details so that they will align vertically and horizontally with details of a similar system when they are printed together on one sheet. Convenient note and arrow placement should also be considered.

Notes will be typed beginning 14/8ths (1¾″) from the right-hand edge of the detail. The drawing itself should end at least ¼″ from this typing margin.

The circle for the grid number is ½″ in diameter and is placed in the lower left hand corner of the detail, ¼″ from both edges.

If dimensions are included in a detail, they should be positioned on the left side and bottom so as not to be confused with the typewritten notes and arrows on the right.

Figure 3-6 Standardized Detail Drawing Format (Clark & Van Voorhis Architects, Inc., Phoenix, Arizona).

▷ FREEHAND OR MECHANICALLY DRAWN ON STANDARD SIZE SHEETS OF PAPER OR MYLAR.

Standardized detail sheets are prepared for filing.

Modular grid sheets are preprinted for layout of final working drawing sheets.

▷ PHOTOGRAPHS

Pictures taken of actual construction detail in the field.

Photographs of actual products, materials, and details shown in catalogs and magazines.

▷ PHOTOCOPY

Copies made of existing details selected from historic data bank or previous working drawings.

By photocopying details unto film for application on final working drawing.

▷ STICK ON DETAILS

Copies of details are made on an adhesive backed film and applied to working drawing sheet.

▷ MICROFILM

Details are processed for microfilm storage or aperture cards.

Copies and printout of actual detail can be made for final working drawing.

▷ COMPUTER

Detail is input to computer storage by drawing with the digitizer or keyboard. Upon completion, the computer can be programmed to have plotter prepare a completed working drawing or it can have printer make a copy of one detail.

Figure 3-7 Methods of Preparing Details.

▷ Linework
▷ Lettering and dimensioning
▷ Symbols
▷ Abbreviations
▷ Terminology
▷ Scales
▷ Sheet sizes
▷ Notes
▷ Numbering

3-4.4 Develop an office standards manual.

Specific procedures and guidelines should be outlined in an office procedures manual for use by members of the design team. It should include an agreed on format for organization and development of information accessed by each person responsible for developing standard details. Training sessions should also be conducted so that new members of the design team understand the procedures and requirements for detail banking. Several important sections of an office procedures manual are shown in Figure 3-14.

The communication methods presented in Figures 3-9 and 3-10 show how a firm organizes and controls the use of symbols and abbreviations in preparing contract documents.

The diagrams in Figure 3-11 show how door and hardware nomenclature can be standardized according to established criteria.

Comprehensive drafting standards shown in Figures 3-12 and 3-13 were selected from two major reference sources that outline procedures for standardizing working drawing communication [6], [7].

3-4.5 Identify steps and procedures to be followed in developing standard details for banking.

To ensure that details are prepared in a standard format, procedures, organizational requirements, and developmental methods must all be identified. Identifying critical steps in a design development process will enable the person in charge of detail banking to administer and monitor the overall system effectiveness. Coordination will be easier to attain through use of a procedural process that is keyed to specific activities required to achieve fulfillment of the banking system. The critical steps in detail preparation are the following:

1. Select a standard size detail sheet—$8\frac{1}{2} \times 11$ inch is used by many firms.

2. Use the most effective medium for preparing details—research graphics industry for best available product.

3. Organize information in an acceptable standard format—determine the relationship between detail elements.

4. Develop a uniform and consistent placement for notation—study and evaluate visual clarity of information.

5. Select a standard lettering style and size—legibility after reproduction is critical for communication.

6. Use standard format for dimensioning, line weights, and arrows—continuity and legibility will improve understanding.

7. Standardize all symbols used in details—clear and precise representation conveys meaningful information.

3-4.6 Select and develop details that show critical construction relationships.

Selecting critical details becomes an important step in the design development process. A mock-up of construction drawings showing details of the project should be prepared before pursuing specific detail design. Irrelevant details and unclear section cuts should be eliminated. Only clear, precise, and meaningful details should be selected for a final working drawing. Highly abstract details will help to confuse the contractor during construction. Therefore it is essential that

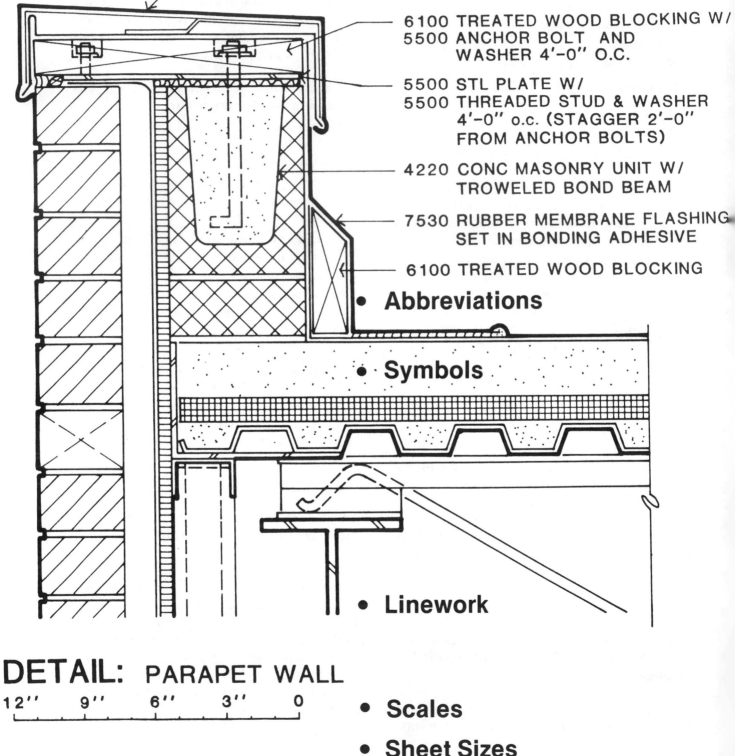

- **Notes**
- **Lettering**
- **Terminology**

7800 ALUM PARAPET CAP

6100 TREATED WOOD BLOCKING W/
5500 ANCHOR BOLT AND
WASHER 4'-0" O.C.

5500 STL PLATE W/
5500 THREADED STUD & WASHER
4'-0" o.c. (STAGGER 2'-0"
FROM ANCHOR BOLTS)

4220 CONC MASONRY UNIT W/
TROWELED BOND BEAM

7530 RUBBER MEMBRANE FLASHING
SET IN BONDING ADHESIVE

6100 TREATED WOOD BLOCKING

- **Abbreviations**
- **Symbols**
- **Linework**

DETAIL: PARAPET WALL

12''   9''   6''   3''   0

- **Scales**
- **Sheet Sizes**

Figure 3-8 Standardizing the Elements of Communication. (Detail Content: Gresham, Smith and Partners, Nashville, Tennessee.)

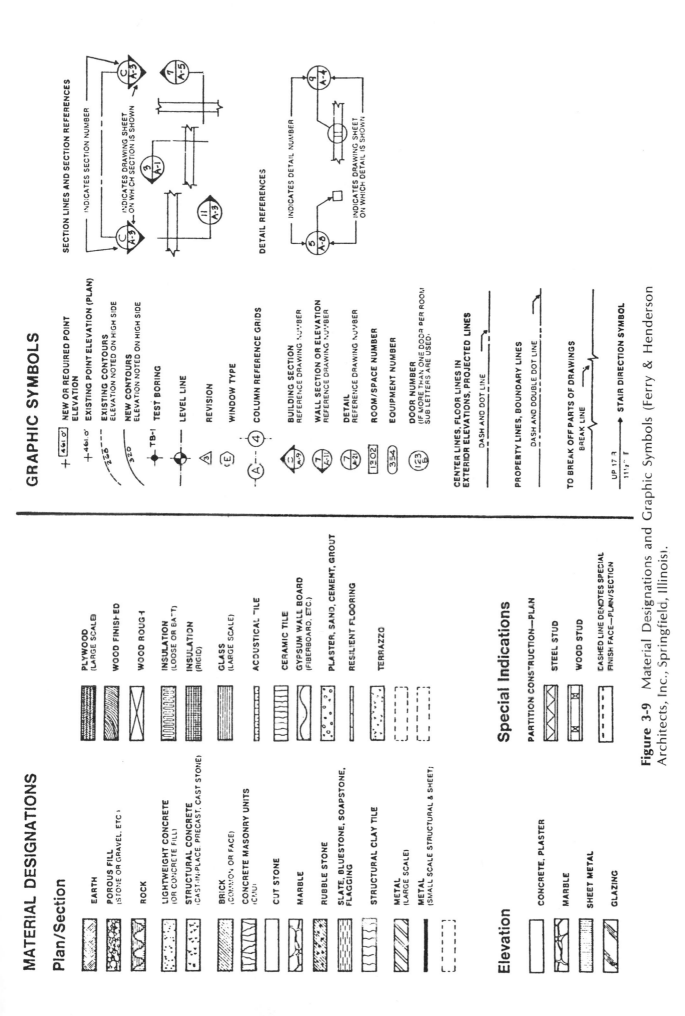

Figure 3-9 Material Designations and Graphic Symbols (Ferry & Henderson Architects, Inc., Springfield, Illinois).

ARCHITECTURAL WORKING DRAWING ABBREVIATIONS

SYMBOLS used as abbreviations:

L	angle
℄	centerline
c	channel
d	penny
⊥	perpendicular
PL	plate
⌀	round

ABBREVIATIONS:

ABV	above
AFF	above finished floor
ASC	above suspended ceiling
ACC	access
ACFL	access floor
AP	access panel
AC	acoustical
ACPL	acoustical plaster
ACT	acoustical tile
ACR	acrylic plastic
ADD	addendum
ADH	adhesive
ADJ	adjacent
ADJT	adjustable
AGG	aggregate
A/C	air conditioning
ALT	alternate
AL	aluminum
ANC	anchor, anchorage
AB	anchor bolt
ANOD	anodized
APX	approximate
ARCH	architect (ural)
AD	area drain
ASB	asbestos
ASPH	asphalt
AT	asphalt tile
AUTO	automatic
BP	back plaster (ed)
BSMT	basement
BRG	bearing
BPL	bearing plate
BJT	bed joint
BM	bench mark
BEL	below
BET	between
BVL	beveled
BIT	bituminous
BLK	block
BLKG	blocking
BD	board
BS	both sides
BW	both ways

BOT	bottom
BRK	brick
BRZ	bronze
BLDG	building
BUR	built up roofing
BBD	bulletin board
CAB	cabinet
CAD	cadmium
CPT	carpet (ed)
CSMT	casement
CI	cast iron
CIPC	cast-in-place concrete
CST	cast stone
CB	catch basin
CK	calk (ing) caulk (ing)
CLG	ceiling
CHT	ceiling height
CEM	cement
PCPL	cement plaster (portland)
CM	centimeter(s)
CER	ceramic
CT	ceramic tile
CMT	ceramic mosaic (tile)
CHBD	chalkboard
CHAM	chamfer
CR	chromium (plated)
CIR	circle
CIRC	circumference
CLR	clear (ance)
CLS	closure
COL	column
COMB	combination
COMPT	compartment
COMPO	composition (composite)
COMP	compress (ed), (ion), (ible)
CONC	concrete
CMU	concrete masonry unit
CX	connection
CONST	construction
CONT	continuous or continue
CONTR	contract (or)
CLL	contract limit line
CJT	control joint
CPR	copper
CG	corner guard
CORR	corrugated
CTR	counter
CFL	counterflashing
CS	countersink
CTSK	countersunk screw
CRS	course (s)
CRG	cross grain
CFT	cubic foot
CYD	cubic yard

DPR	damper
DP	dampproofing
DL	dead load
DEM	demolish, demolition
DMT	demountable
DEP	depressed
DTL	detail
DIAG	diagonal
DIAM	diameter
DIM	dimension
DPR	dispenser
DIV	division
DR	door
DA	doubleacting
DH	double hung
DTA	dovetail anchor
DTS	dovetail anchor slot
DS	downspout
D	drain
DRB	drainboard
DT	drain tile
DWR	drawer
DWG	drawing
DF	drinking fountain
DW	dumbwaiter
EF	each face
E	east
ELEC	electric (al)
EP	electrical panelboard
EWC	electric water cooler
EL	elevation
ELEV	elevator
EMER	emergency
ENC	enclose (ure)
EQ	equal
EQP	equipment
ESC	escalator
EST	estimate
EXCA	excavate
EXH	exhaust
EXG	existing
EXMP	expanded metal plate
EB	expansion bolt
EXP	exposed
EXT	exterior
EXS	extra strong
FB	face brick
FOC	face of concrete
FOF	face of finish
FOM	face of masonry
FOS	face of studs
FF	factory finish
FAS	fasten, fastener
FN	fence

Figure 3-10 Architectural Working Drawing Abbreviations (Ferry & Henderson Architects, Inc., Springfield, Illinois).

FBD	fiberboard	**HWD**	hardwood	**MRB**	marble
FGL	fiberglass	**HJT**	head joint	**MAS**	masonry
FIN	finish (ed)	**HDR**	header	**MO**	masonry opening
FFE	finished floor elevation	**HTG**	heating	**MTL**	material (s)
FFL	finished floor line	**HVAC**	heating/ventilating/air	**MAX**	maximum
FA	fire alarm		conditioning	**MECH**	mechanic (al)
FBRK	fire brick	**HD**	heavy duty	**MC**	medicine cabinet
FE	fire extinguisher	**HT**	height	**MED**	medium
FEC	fire extinguisher cabinet	**HX**	hexagonal	**MBR**	member
FHS	fire hose station	**HES**	high early-strength cement	**MMB**	membrane
FPL	fireplace	**HC**	hollow core	**MET**	metal
FP	fireproof	**HM**	hollow metal	**MFD**	metal floor decking
FRC	fire-resistant coating	**HK**	hook (s)	**MTFR**	metal furring
FRT	fire-retardant	**HOR**	horizontal	**MRD**	metal roof decking
FLG	flashing	**HB**	hose bibb	**MTHR**	metal threshold
FHMS	flathead machine screw	**HWH**	hot water heater	**M**	meter (s)
FHWS	flathead wood screw			**MM**	millimeter (s)
FLX	flexible	**INCIN**	incinerator	**MWK**	millwork
FLR	floor (ing)	**INCL**	include (d), (ing)	**MIN**	minimum
FLCO	floor cleanout	**ID**	inside diameter	**MIR**	mirror
FD	floor drain	**INS**	insulate (d), (ion)	**MISC**	miscellaneous
FPL	floor plate	**INSC**	insulating concrete	**MOD**	modular
FLUR	fluorescent	**INSF**	insulating fill	**MLD**	molding, moulding
FJT	flush joint	**INT**	interior	**MR**	mop receptor
FTG	footing	**ILK**	interlock	**MT**	mount (ed), (ing)
FRG	forged	**INTM**	intermediate	**MOV**	movable
FND	foundation	**INV**	invert	**MULL**	mullion
FR	frame (d), (ing)	**IPS**	iron pipe size		
FRA	fresh air	**JC**	janitor's closet	**NL**	nailable
FS	full size	**JT**	joint	**NAT**	natural
FBO	furnished by others	**JF**	joint filler	**NI**	nickel
FUR	furred (ing)	**J**	joist	**NR**	noise reduction
FUT	future			**NRC**	noise reduction
		KCPL	Keene's cement plaster		coefficient
GA	gage, gauge	**KPL**	kickplate	**NOM**	nominal
GV	galvanized	**KIT**	kitchen	**NMT**	nonmetallic
GI	galvanized iron	**KO**	knockout	**N**	north
GP	galvanized pipe			**NIC**	not in contract
GSS	galvanized steel sheet	**LBL**	label	**NTS**	not to scale
GKT	gasket (ed)	**LAB**	laboratory		
GC	general contract (or)	**LAD**	ladder	**OBS**	obscure
GL	glass, glazing	**LB**	lag bolt	**OC**	on center (s)
GLB	glass block	**LAM**	laminate (d)	**OP**	opaque
GLF	glass fiber	**LAV**	lavatory	**OPG**	opening
GCMU	glazed concrete masonry	**LH**	left hand	**OJ**	open-web joist
	units	**L**	length	**OPP**	opposite
GST	glazed structural tile	**LT**	light	**OPH**	opposite hand
GB	grab bar	**LC**	light control	**OPS**	opposite surface
GD	grade, grading	**LP**	lightproof	**OD**	outside diameter
GRN	granite	**LW**	lightweight	**OHMS**	ovalhead machine screw
GVL	gravel	**LWC**	lightweight concrete	**OHWS**	ovalhead wood screw
GF	ground face	**LMS**	limestone	**OA**	overall
GT	grout	**LTL**	lintel	**OH**	overhead
GPDW	gypsum dry wall	**LL**	live load		
GPL	gypsum lath	**LVR**	louver	**PNT**	paint (ed)
GPPL	gypsum plaster	**LPT**	low point	**PNL**	panel
GPT	gypsum tile			**PB**	panic bar
		MB	machine bolt	**PTD**	paper towel dispenser
HH	handhole	**MI**	malleable iron	**PTR**	paper towel receptor
HBD	hardboard	**MH**	manhole	**PAR**	parallel
HDW	hardware	**MFR**	manufacture (er)	**PK**	parking

Figure 3-10 *(Continued)*

PBD	particle board	**RM**	room	**TOL**	tolerance
PTN	partition	**RO**	rough opening	**T&G**	tongue and groove
PV	pave (d), (ing)	**RB**	rubber base	**TSL**	top of slab
PVMT	pavement	**RBT**	rubber tile	**TST**	top of steel
PED	pedestal	**RBL**	rubble stone	**TW**	top of wall
PERF	perforate (d)			**TB**	towel bar
PERI	perimeter	**SFGL**	safety glass	**TR**	transom
PLAS	plaster	**SCH**	schedule	**T**	tread
PLAM	plastic laminate	**SCN**	screen	**TYP**	typical
PL	plate	**SNT**	sealant		
PG	plate glass	**STG**	seating	**UC**	undercut
PWD	plywood	**SEC**	section	**UNF**	unfinished
PT	point	**SSK**	service sink	**UR**	urinal
PVC	polyvinyl chloride	**SHTH**	sheathing		
PE	porcelain enamel	**SHT**	sheet	**VJ**	v-joint (ed)
PTC	post-tensioned concrete	**SG**	sheet glass	**VB**	vapor barrier
PCF	pounds per cubic foot	**SH**	shelf, shelving	**VAR**	varnish
PFL	pounds per lineal foot	**SHO**	shore (d), (ing)	**VNR**	veneer
PSF	pounds per square foot	**SIM**	similar	**VRM**	vermiculite
PSI	pounds per square inch	**SKL**	skylight	**VERT**	vertical
PCC	precast concrete	**SL**	sleeve	**VG**	vertical grain
PFB	prefabricate (d)	**SC**	solid core	**VIN**	vinyl
PFN	prefinished	**SP**	soundproof	**VAT**	vinyl asbestos tile
PRF	preformed	**S**	south	**VB**	vinyl base
PSC	prestressed concrete	**SPC**	spacer	**VF**	vinyl fabric
PL	property line	**SPK**	speaker	**VT**	vinyl tile
		SPL	special		
QT	quarry tile	**SPEC**	specification (s)	**WSCT**	wainscot
		SQ	square	**WTW**	wall to wall
RBT	rabbet, rebate	**SST**	stainless steel	**WH**	wall hung
RAD	radius	**STD**	standard	**WC**	water closet
RL	rail (ing)	**STA**	station	**WP**	waterproofing
RWC	rainwater conductor	**ST**	steel	**WR**	water repellent
REF	reference	**STO**	storage	**WS**	waterstop
RFL	reflect (ed), (ive), (or)	**SD**	storm drain	**WWF**	welded wire fabric
REFR	refrigerator	**STR**	structural	**W**	west
REG	register	**SCT**	structural clay tile	**WHB**	wheel bumper
RE	reinforce (d), (ing)	**SUS**	suspended	**W**	width, wide
RCP	reinforced concrete pipe	**SYM**	symmetry (ical)	**WIN**	window
REM	remove	**SYN**	synthetic	**WG**	wired glass
RES	resilient	**SYS**	system	**WM**	wire mesh
RET	return			**WO**	without
RA	return air	**TKBD**	tackboard	**WD**	wood
RVS	reverse (side)	**TKS**	tackstrip	**WB**	woodbase
REV	revision (s), revised	**TEL**	telephone	**WPT**	working point
RH	right hand	**TV**	television	**WI**	wrought iron
ROW	right of way	**TC**	terra cotta		
R	riser	**TZ**	terrazzo	**EPNT**	epoxy paint
RVT	rivet	**THK**	thick (ness)	**PR**	pair
RD	roof drain	**THR**	threshold	**SV**	sheet vinyl
RFH	roof hatch	**TPTN**	toilet partition	**SGL**	single
RFG	roofing	**TPD**	toilet paper dispenser	**TG**	tempered glass
				STN	stained

Figure 3-10 *(Continued)*

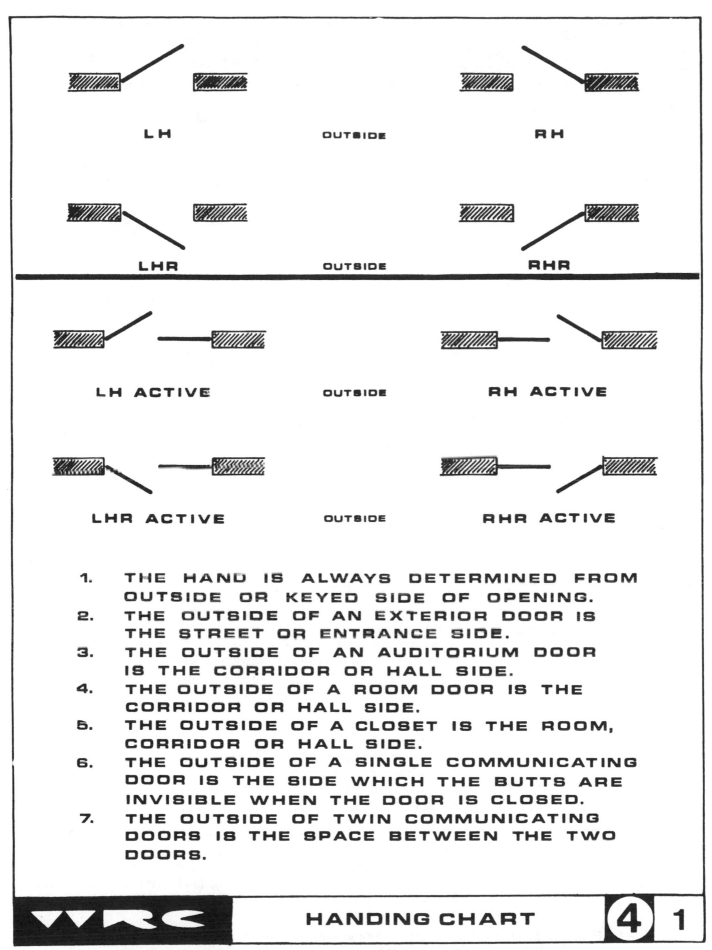

Figure 3-11 Door Hardware Handing Chart (R. C. Hollow Metal Company, Denver, Colorado).

DRAWING REFERENCE SYMBOL

8 - drawing number (on sheet A5)
A5 - sheet number on which drawing is presented (architectural sheet number 5)
8-A5 - indication for drawing number and sheet number on which drawing is presented, in notes, schedules, and specifications.

DOOR REFERENCE SYMBOL - SCHEDULE METHOD

2 - floor number
07 - door opening (on second floor)
207 - door number
See list of drawings for sheet numbers on which door schedules, door types, and door frame details are presented.

DOOR REFERENCE SYMBOL-DIRECT REFERENCE METHOD

D - door type
6 - door frame detail number
b - hardware group letter
See list of drawings for sheet numbers on which door types and door frame details are presented. See specifications for contents of hardware group letter.

WINDOW REFERENCE SYMBOL

B - window type
3 - window detail number
See list of drawings for sheet numbers on which window types and window details are presented.

TITLE REFERENCE SYMBOL

Drawing titles are placed under all drawings, except some building component type drawings, such as door type and window type drawings. Locations of details on building sites or in buildings are sometimes included in detail drawing titles for general referencing purposes only. Do not use information in drawing titles to determine numbers or quantities.

D - door type
6 - door frame detail number*
B - window type
3 - window detail number*
8 - drawing number (on sheet A5)*
A5 - sheet number on which drawing is presented (architectural sheet number 5)

*In some grouped detail drawings, especially door frame detail drawings and window detail drawings, drawing numbers may be suffixed by letters to differentiate detail drawings. For example, window details consisting of head, sill, jamb, vertical mullion, and horizontal mullion detail drawings might be designated 3A, 3B, 3C, 3D, and 3E respectively.

Figure 3-12 Office Architectural Drafting Standards and Symbols (O'Connell, *Graphic Communications in Architecture.*)

DRAFTING STANDARDS & TECHNIQUES

DRAFTING TECHNIQUES

MICROFONT
ABCDEFGHIJKLMNOPQR
STUVWXYZ 1234567890

THIS IS AN EXAMPLE OF 5/32" MICROFONT. IT WAS ESPECIALLY CREATED BY THE NATIONAL MICROFILM ASSOCIATION'S STANDARDS COMMITTEES.

	Actual Size of Lettering	¾ Size Reduction	⅔ Size Reduction	½ Size Reduction	⅓ Size Reduction	¼ Size Reduction
.350-inch	ABCD abcdef 12345	ABCD abcdef 12345	ABCD abcdef 12345	ABCD abcdef 12345	ABCD abcdef 12345	ABCD abcdef 12345
.240-inch	ABCDEF abcdefgh 1234567	ABCDEF abcdefgh 1234567	ABCDEF abcdefgh 1234567	ABCDEF abcdefgh 1234567	ABCDEF abcdefgh 1234567	ABCDEF abcdefgh 1234567
.140-inch	ABCDEFGHIJK abcdefghijklmno 1234567890°."	ABCDEFGHIJK abcdefghijklmno 1234567890°."	ABCDEFGHIJK abcdefghijklmno 1234567890°."	ABCDEFGHIJK abcdefghijklmno 1234567890°."	ABCDEFGHIJK abcdefghijklmno 1234567890°."	ABCDEFGHIJK abcdefghijklmno 1234567890°."
.090-inch	ABCDEFGHIJKLMNOP 1234567890°."= % ()	ABCDEFGHIJKLMNOP 1234567890°."= % ()	ABCDEFGHIJKLMNOP 1234567890°."= % ()	ABCDEFGHIJKLMNOP 1234567890°."= % ()	ABCDEFGHIJKLMNOP 1234567890°."= % ()	ABCDEFGHIJKLMNOP 1234567890°."= % ()

COMMITTEE ON PRODUCTION OFFICE PROCEDURES

©JULY 1980
NORTHERN CALIFORNIA CHAPTER
AMERICAN INSTITUTE OF ARCHITECTS

Figure 3-13 Drafting Standards and Techniques (Northern California Chapter of the American Institute of Architects, San Francisco, California).

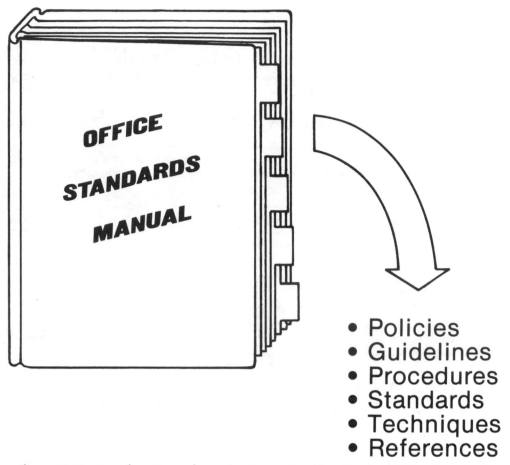

- Policies
- Guidelines
- Procedures
- Standards
- Techniques
- References

Figure 3-14 Develop Procedures for Preparing Details: the Office Standards Manual.

informative details be carefully selected for detail banking because they will have the greatest potential for reuse and approval during field evaluation. The mock-up drawings in Figures 3-15, 3-16, 3-17, and 3-18 demonstrate the preplanning necessary to identify critical detail relationships.

3-4.7 Standardizing detail notation.

Detail notation can be standardized and identified with specification sections by referencing the number of each applicable section. Steps to developing standard notation are listed as follows:

1. Select descriptors from a construction thesaurus to generate a language for notation on details.

2. Structure reusable word phases that define the elements of each detail.

3. Prepare standard notation that can be indexed, stored, and retrieved.

4. Store notation by descriptor term identification.

5. Correlate the specification section numbers with notation for ease in cross-referencing.

6. Select a method for processing the prepared notation, manually, by word processor, or by computer.

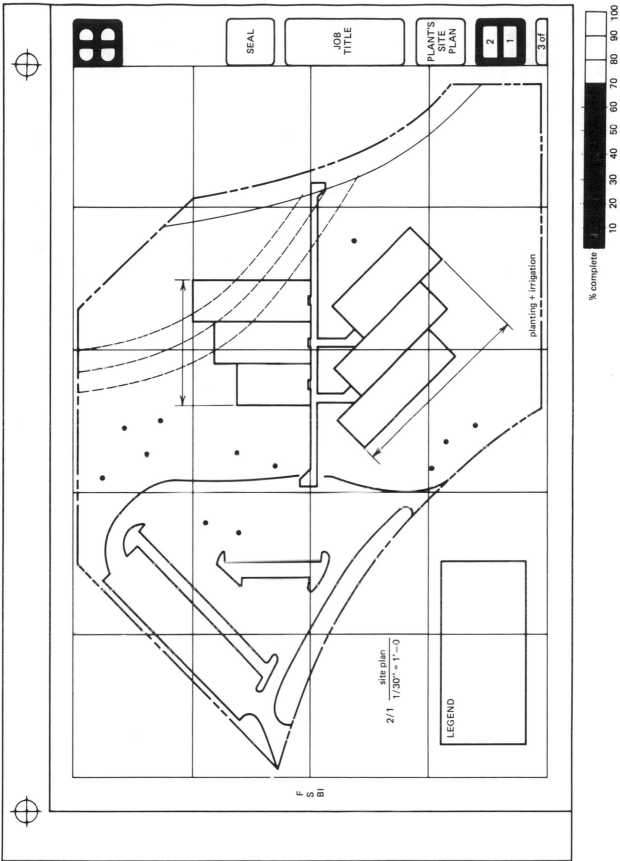

Figure 3-15 Mock-Up Site Plan (Bullock/Graves & Associates, Architects Planners, Inc., Pensacola, Florida).

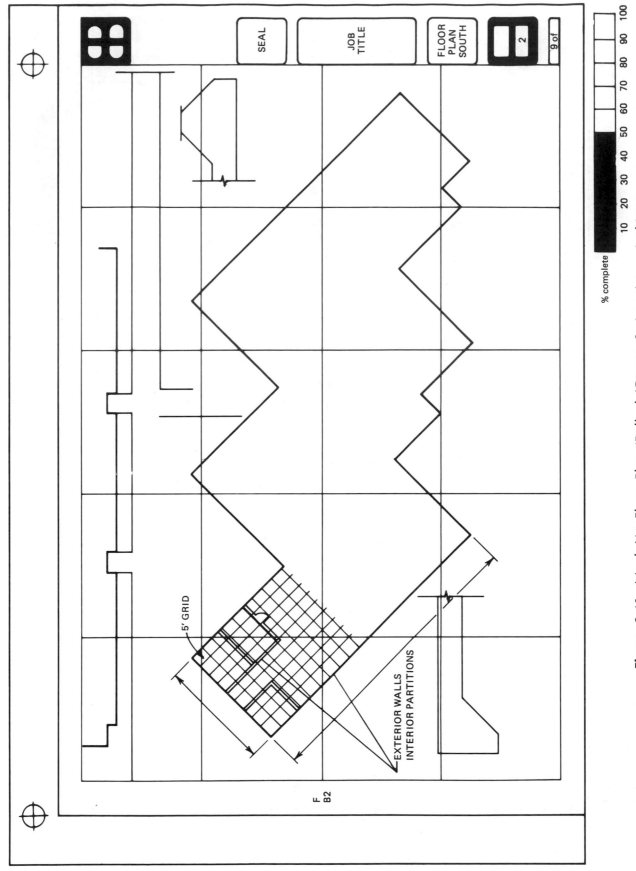

Figure 3-16 Mock-Up Floor Plan (Bullock/Graves & Associates, Architects Planners, Inc. Pensacola, Florida).

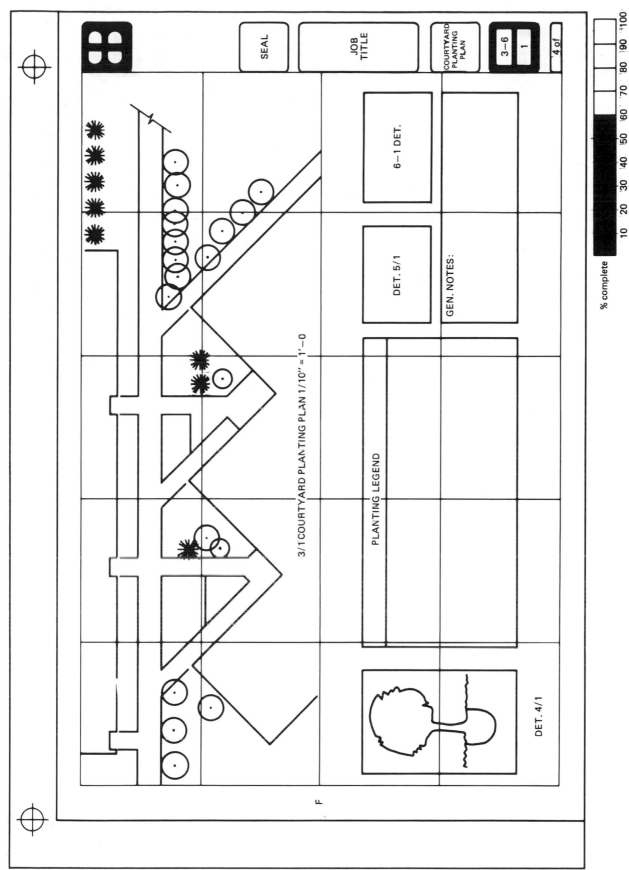

Figure 3-17 Mock-Up Planting Plan (Bullock/Graves & Associates, Architects Planners, Inc., Pensacola, Florida).

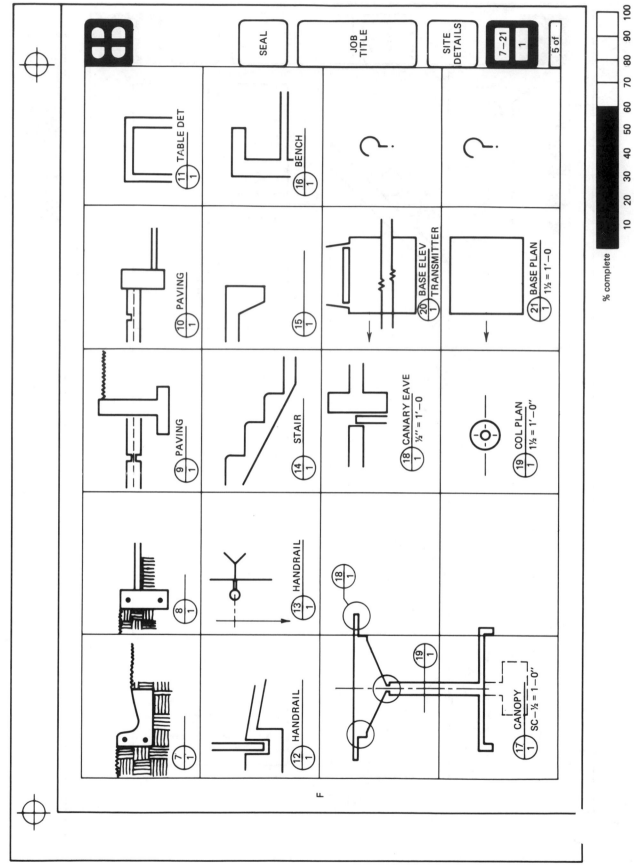

Figure 3-18 Mock-Up Detail Sheet (Bullock/Graves & Associates, Architects Planners, Inc., Pensacola, Florida).

Example Creating Standard Notation:

▷ <u>Stair Handrails</u> mounted to wall at 4–0″ O/C.

▷ <u>Acoustic Insulation</u> placed in all openings above dropped ceilings.

▷ <u>Illumination Levels</u> controlled by <u>Flexible Lighting Design</u>

▷ Specification number reference (see detail in Figure 3-8, Spec. No. 7800 <u>Aluminum Parapet Cap</u>).

Underlined descriptors are selected from a construction language and used in creating standard notation. These notes can be standardized and stored in word processors and computers for use on future projects. Terminology control is established by utilizing descriptors approved in the construction language. Working drawings and specifications can be language coordinated through use of standard notation and approved descriptors. Chapter 6 demonstrates how an interactive data base can provide access to appropriate specification sections.

3-5 QUALITY CONTROL DURING DETAIL DEVELOPMENT

An office anticipating development of a detail banking system should also establish a quality control program. To manage a large work force, mass production, and a rapid turnover of personnel and projects, it is necessary to develop effective control procedures. Design development and field construction control must be established before starting a project. High-quality details can only be derived by thorough research, study, and dedicated design input. Each step of this process must be monitored in accordance with an identified plan of action that includes critical checkpoints to measure the quality of details. A good quality control program will enable the designer to detect and solve problems before they occur

during design development, production, and construction.

During the design development stage the materials and products being considered for construction must be carefully evaluated. Research findings, tests, and other criteria must be gathered to evaluate proposed material relationships. Checkpoints and evaluation procedures should be systematically structured to review the detail design and construction process. (The form presented in Figure 3-22 will enable the designer to document the evaluation process associated with each detail selected for banking).

Development of a quality control program requires a thorough analysis of office goals and objectives. Quality control procedures must be carried out toward a defined purpose. Only if a purpose is identified early in the development stage can an objective and directed quality control program evolve. Each step in a quality control program must fulfill an identified need. Steps must be clear, simple, and easy to implement during the design development stage. The plan must be structured so that all design professionals responsible for detail development have a clear understanding of its goals and purpose. Each organization must define its own plan of action so that the resulting quality control program relates specifically to its intended goal.

Schedules, checklists, and guidelines should be prepared so that each member of the design team knows the critical coordination checkpoints during the design process. Checklists by themselves will not necessarily improve a construction detail. However, by using the checklist, the design professional can be assured that proper procedures for developing the details were carried out to the best of one's ability. For the young professional a checklist serves as an excellent guide to fulfilling important steps systematically. It will also serve as a record of procedures for the design professional in the event that legal problems occur. To implement a successful quality control program, it is important to develop a number of quality control procedures.

3-6 DEVELOPING QUALITY CONTROL PROCEDURES

3-6.1 Establish a systematic process for detail development.

An effective quality control program must start at the very first stage of detail development. Steps to monitor the planning, development, and design methods are critical in maintaining the level of uniformity required in construction communication. A format and system for checking each stage of development is critical in structuring a successful detail banking program. One individual, highly experienced in construction technology and materials, should be assigned the responsibility for monitoring the development process.

3-6.2 Prepare standards, guidelines, and checklists for the design and production team.

Detail monitoring must involve all members of the design team. Each member should have a copy of the standards, guidelines, and checklists to be used during the detail development process. Specific checkpoints should be highlighted so that each step is carried out in accordance with the overall plan of action.

3-6.3 Identify checkpoints for detail review, evaluation, and approval.

It is important to identify specific checkpoints for each step in the monitoring process to ensure that checklists are fulfilled in accordance to the quality control program. Specific relationships between drawings and specifications should be checked for terminology and accuracy. Overall design consistency should be evaluated during early design stages.

3-6.4 Develop field evaluation procedures to provide detail feedback information.

An effective quality control program must be coordinated with all field construction procedures and provide a framework for feedback. In order to achieve this goal, a field inspection record sheet must be developed, completed, and distributed to the design team. The record sheet should maintain all field input regarding problems, inconsistencies, and errors in communication that may have occurred as a result of the poor details. All detail construction should be evaluated and documented on the field evaluation form. Records should be maintained and logged according to the disposition identified for a specific detail. A record file, for future reference, should be maintained in the field office as well as in the drafting room.

3-6.5 Establish detail performance requirements before production.

To enable effective quality control procedures to evolve, it is important that detail performance requirements be identified early in the design stage. These requirements should be stated in a manner that will achieve the client's anticipated needs. Without specific information on detail performance requirements it is difficult to establish effective quality control procedures. Each quality control measure should be aligned with a specific requirement. This is the most essential part of a good quality control program.

3-6.6 Request all technical data and performance information from manufacturers of products and materials.

Information identifying the proper procedures for use and application of all products should be obtained from manufacturers before design development. A systematic selection and evaluation process should take place before detail development is undertaken. Quality control procedures should be applied to evaluation and selection of products and materials. Specifications should also be obtained and evaluated as a part of the monitoring process. The material and product data form shown in Figure 2-8 should record all decisions regarding selection and application of a specific manufacturer's product. All data obtained should be filed under the detail numbering system so that cross-referencing can be maintained between each detail and its approved materials.

3-6.7 Establish a monitoring program that tracks details from development to construction.

Quality control procedures must be established for each step in the design development and construction process. Timing for all checkpoints should be identified and recorded in accordance with the overall quality control program. Identify team members responsible for carrying out each step in the monitoring process. Record sheets should be maintained identifying the item checked, the time of checking, and who was responsible for carrying out that check. Figure 3-19 indicates steps and procedures recommended in quality control [5].

3-6.8 Identify steps and timing required to coordinate review, evaluation, and approval of each detail during the development process.

Specific steps to check adequately detail development are essential in the monitoring program. Checkpoint timing must be identified to accomplish the monitoring task during the critical development stage. An early review schedule will ensure that a detail fulfills established requirements set forth in the design stage.

3-6.9 Assign personnel to administer and monitor the detail development process.

A successful quality control program requires good management of critical task assignments. Good management programs have someone appointed to control the overall monitoring process, that is, to ensure that specific steps are carried out in accordance with the overall plan. At each level of the project's development key people are assigned to coordinate assignments with the overall administrator of the program. Procedural manuals help to ensure that each person knows and understands his or her responsibilities.

3-6.10 Record all problems and approvals for future reference.

Record forms should be completed and filed for future reference. These forms should contain all information regarding problems, decisions, and approvals associated with each detail developed. All record sheets shall be maintained and filed in accordance with the detail numbering system.

The highlights of good quality control have been identified as a framework for structuring a successful program. Examples of quality control procedures in a design professional's office are shown in Figures 3-20 and 3-21 for existing and workable programs. Quality control measures carried out in related disciplines should be researched prior to developing an in-house program. However, it is important that quality control techniques be structured in relation to the communication technology capable of monitoring a system, namely, word processors and computers.

RECOMMENDATIONS

- ■ **Insist upon neat, legible drawings.**

- ■ **Implement an office manual of working drawing standards and procedures.**

- ■ **Maintain open and defined channels of communication among all personnel involved in the project.**

- ■ **Coordinate drawings by engineering consultants.**

- ■ **Arrange drawings in a logical sequence.**

- ■ **Remember to coordinate the drawings with the specifications avoiding conflicts, contradictions or ambiguities.**

- ■ **Insist upon the review of all drawings by a principal or qualified supervisor.**

Figure 3-19 Quality Control in the Preparation of Working Drawings. (*Guidelines for Improving Practice—Architects and Engineers Professional Liability,* vol. 1, no. 2. Victor O. Schinnerer & Company, Inc., © 1971.)

A Quality Control Program*

A. Design documentation review (at the end of design development).
Objectives:

1. Review adequacy and completeness of design documentation.
2. Review proposed design and detailing with respect to technical ramifications.
3. Review of zoning and code compliance.
4. Review of objectives, procedures, and requirements for construction document reviews.

*Source: Perkins & Will, Inc., Chicago, Illinois.

Procedure:

1. Coordinating Job Captain informally reviews requirements with Project Director, Job Captain, and other team members as appropriate and reports findings to Director of Architectural Services and/or Architectural Technical Services Coordinator.

Requirements:

1. Complete set of design drawings; outline spec, cost estimate, room data sheets, code analysis, and all other design information including survey and soil test.

B. Contract document organization review (before 10 percent completion of contract document).

Objectives:

1. Assures logical and detailed organization of job consistent with firm's objectives.
2. Develops logical sequence for executing the work.
3. Identifies and resolves problem areas.

Procedure:

1. Coordinating Job Captain informally reviews all requirements with Job Captain.
2. Next formal review meeting with Project Director, Project Engineer, Job Captain, Architectural Technical Services Coordinator, and Leading Coordinating Job Captain assures agreement on direction.

Requirements:

1. Schedule of drawings.
2. Composition of drawings.
3. Office/job standards.
4. Format for room finish, door schedule, wall types, equipment, etc.
5. Review of any innovative presentation techniques.
6. Manpower projections.
7. Project schedule.
8. Technique for evaluation of percentage of completion at pay periods.
9. Work force updates at pay periods.
10. Efficiency check (percent complete vs. percent time elapsed vs. percent budget used (worker hours)).
11. Project work force records on per sheet basis.
12. Work authorization.

C. Contract document interim observations.
Objectives:

1. Observe progress of documentation and organization.

2. Observe quality of documentation.
3. Assist with resolution of problems.
4. Remain available for technical discussions.

Procedure:

1. Leading Coordinating Job Captain continues to review informally progress of work and report to Architectural Technical Services Coordinator as appropriate.
2. Periodic review of check sets made by Job Captain.

D. Contract document prebid review (approximately 90 percent completion of contract document).
Objectives:

1. Detailed review of documentation.

Procedure:

1. Informal review briefing, consisting of independent and combined reviews involving Leading Job Captain, Quality Control Engineer, Job Captain, Project Engineer, Designer, and Senior Field Architect.
2. Formal review meeting with Leading Job Captain, Quality Control Engineer, Job Captain, Senior Field Architect, team members as appropriate, Director of Architectural Services, Director of Engineering Services and Architectural Technical Services Coordinator with resolution of actions required, if any.

Requirements:

1. Complete set of drawings (architectural, structural, mechanical, plumbing, and electrical), copy of final specs, and all pertinent project records.

E. Bidding phase.

Objectives:

1. Maintain design and technical integrity of the construction documents.

Procedure:

1. Available for project meetings as required.
2. Monitors written addenda and related drawings.

F. Construction phase.
    Objectives:

    1. Maintain design and technical integrity of construction documents.
    2. Modifications to the contract monitors.

Procedure:

1. Available for project meetings as required.
2. Monitor change orders and other related correspondence.

                                                            □

3-7 FIELD EVALUATION PROCEDURES

Problems occurring in graphic communication of material relationships can be reduced if monitoring procedures are established early in the construction process and in conjunction with design development. Records documenting field problems provide the design professional with a base for future improvement of construction details and working drawings. Files containing detail problem reports serve to provide a reference base for educating the design team and to improve on future construction and design processes. The educational benefits of field feedback are as important as improving future construction details. Feedback systems provide the design professional with the least-expensive means for in-house education and construction improvement. Banking details for future evaluation and use is like putting money in the bank. The "interest" is in the form of improved working drawings and reduced building failures.

In banking construction details, it is necessary to have a field evaluation plan. The plan should concentrate on monitoring details during the construction stage. As part of the monitoring process, documentation must be made to identify problem areas and record accurate field information. Information should be categorized for easy inspection, evaluation, and final review of each detail.

No detail should be left unevaluated during field construction. Field representatives should be required to record problems identified during the construction process. Contractor feedback is also essential to development and improvement of each construction detail. Problems in the assembly of various building materials should be identified for reevaluation during working drawing development. The detail development format should be compatible with procedures and implementation plans devised for the construction detail banking system.

In sum, an effective field evaluation program will:

1. Monitor all building details during the construction process.
2. Develop procedures and guidelines for identifying field problems.
3. Record field information on Detail Evaluation Cards.
4. Request field feedback from the field administrator and the contractor.
5. Provide the construction administrator with a structured format to record evaluation data.
6. Collect field evaluation data in a format that is consistent with a detail banking system.
7. Develop field feedback routing system to appropriate members of design team.
8. Record appropriate data on a Detail Data Card and file information for future retrieval.

Quality Assurance Reviews

Project	Phase	Phase Due Date	Date Documents Received	Comments	Reviews Required
					○○○○○
					○○○○○
					○○○○○
					○○○○○
					○○○○○
					○○○○○
					○○○○○
					○○○○○
					○○○○○
					○○○○○
					○○○○○

Key   ○ Doc't Distributed   ⊕ Comments Written   ✸ Follow up Complete

Figure 3-20   Quality Assurance Review (Perkins & Will, Inc., Chicago, Illinois).

Quality Assurance Review Comments

Project _____ Project No. _____

☐ Architectural ☐ Structural Review _____
☐ Site ☐ HVAC Date _____
☐ Civil ☐ CM Reviewer _____
☐ Specifications ☐ Electrical
☐ Field ☐ Plumb & Fire Protect

Item	Drawing No. or Specification Paragraph	Comments	Action

Figure 3-21 Quality Assurance Review Comments (Perkins & Will, Inc., Chicago, Illinois).

54

The project architect, engineer, or job captain should coordinate field evaluation procedures with the field representative selected for the project. Data collection responsibilities should also be designated to the librarian or research specialist responsible for banking construction details. If individual responsibilities are identified and a procedural plan is structured, each member of the team can be made accountable for carrying out an effective field evaluation program.

Caution should be given to making quality control and field feedback procedures so complex that disadvantages and costs outweigh the program benefits. Only essential steps to identify valuable construction information are required in an effective program. Each step should be structured to retrieve usable information. Consider the following issues before developing the feedback process:

1. What is the purpose and goal of each task?
2. Can organizational approval be obtained to establish program?
3. Are staff available to gather information?
4. Is the monitoring process in its simplest form?
5. Will the information requested help improve detail development?

3-8 DEVELOPMENT OF A FIELD EVALUATION DATA CARD

To aid in development of a quality control program, it is important to structure field evaluation procedures that provide effective and timely feedback from the construction process. A system for monitoring problems during field construction is an essential part of the detail banking system. Firsthand information on the level of success a detail achieves during field construction is essential for the production of future details. The feedback program serves a twofold purpose: (1) existing details can be monitored and eval-

uated during construction, and (2) field construction findings can be incorporated into the development of new details.

The Field Evaluation Data Card shown in Figure 3-22 is designed for entering critical information that can be directed toward detail improvement. This card must identify detail, project, and date on which detail was entered into the system. Project identification will help record the number of times a detail has been reused. This will provide effective reference data in the historic file.

Field evaluation data should be recorded on constructibility, material availability, and ease of detail assembly or installation. Field feedback should also provide an overall picture of the construction team's response to the success of individual details.

Problems arising during the construction stage should be clearly identified for future evaluation and detail improvement. To complete the overall review process, the design professional should obtain the owner's approval and evaluation of material relationships, detail aesthetics, and overall performance. Information gathered during the field evaluation process should be made available to all members of the design and construction team, as well as to the manufacturer.

Responsibility for developing the feedback program rests with the project leader and those directly responsible for working drawing production. In planning the field administration program, it is necessary to establish a field-to-office communication system. This system should be planned and programmed so the field administrator can conduct evaluation tasks without interrupting the flow of construction activities. To accomplish this goal, a preplanned Field Evaluation Data Card should be prepared by the office staff and made available to the field administrator.

Each detail selected for evaluation must have its own data card with detail and project identification listed for field personnel. An office prepared notebook of evaluation cards keyed to the working drawings will greatly assist field staff in responding to the feedback program. If the office

FIELD EVALUATION DATA CARD

Bank Detail No.:	
Project Detail No.:	Sheet No.:
Detail Title/Description:	

Detail Status:	New:	Standard:
Prepared By:	Date:	
Project No.:	Date:	
Project Name:	Location:	

Field Evaluator:	Date:

Evaluation Data

Construction Issues		Not Acceptable	Poor	Good	Excellent	Acceptable
1	Material Availability					
2	Constructibility					
3	Installation					
4	Completed Assembly					
5	Performance					
6	Aesthetic Response					
Evaluator's Response						
1	Construction Admin.					
2	Contractor					
3	Manufacturers					
4	Owner					

Field Problems

1	
2	
3	
4	
5	
6	

Data Card Routing

	Staff	Action	Date
	Project Architect		
	Project Engineer		
	Designer		
	Job Captain		
	Draftpersons		
	Consultant		
	Contractor		
	Manufacturers		
	Construction Admin.		

Detail Corrections

Corrections Made:	Approved By	Date
1		
2		
3		

Figure 3-22 Field Evaluation Data Card.

is small, many of the responsibilities for field evaluation will be shared by the production staff. Streamlining the steps and procedures to the evaluation process should be an ongoing program for the project leader.

3-9   BENEFITS OF A FIELD EVALUATION PROGRAM

To carry out field evaluation procedures, it is necessary to follow an identified plan of action. This plan should show the steps necessary to implement an effective system within the framework of the construction administration program. Key people should be appointed to monitor critical steps within the plan. An effective field evaluation plan can provide all involved with a number of benefits.

3-9.1   Compels individuals to be involved in analyzing construction details during the design-development and construction stages.

A structured plan of evaluation will guide individual team members in research and investigation of construction detail development. Valuable field information and material performance data will enable the design team to improve details. A plan of action will stimulate individuals to become more active in the evaluation process. It will also encourage designers to become more aware of potential problems arising out of the design stage.

3-9.2   Requires individuals to document detail information in a systematic and standard way so information can be quickly and easily utilized by other members of the project team.

A formal field evaluation plan will enable detail evaluators to document information in a systematic and organized manner. Critical items should be highlighted so that potential problems can be quickly and easily identified. Proper organization of the information will allow for easy access of critical data during detail design development.

3-9.3   Enables construction detailer to incorporate learning experiences from a greater number of people during the design development and construction review process.

A thorough analysis of detail construction will provide valuable feedback from several disciplines in the construction industry. This information will help the designer check problem areas interfacing different systems. For example, construction of a curtain wall system involves a composite wall system of glass, structural framing, and special connections. Therefore several disciplines must utilize information present in the details. Through an organized systematic review process feedback from each discipline can be gathered and documented in an efficient manner to assist the designer in future detail development.

3-9.4   Provides a means for obtaining immediate feedback on problem areas that develop during project construction.

The systematic field feedback plan will provide a means for obtaining any problem information that arises during the construction process. Daily recording of information can be communicated through an organized feedback plan. Construction administrators should follow specific guidelines for recording field problems. A log book containing field evaluation data cards should be maintained on a daily basis so that detail problems can be reported to production staff immediately.

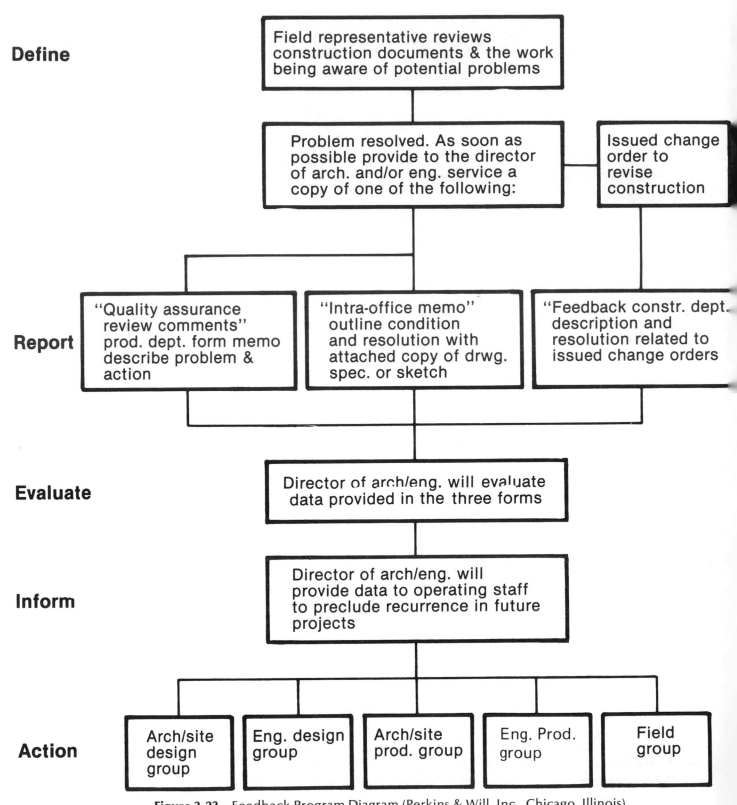

Define

Field representative reviews construction documents & the work being aware of potential problems

Problem resolved. As soon as possible provide to the director of arch. and/or eng. service a copy of one of the following:

Issued change order to revise construction

Report

"Quality assurance review comments" prod. dept. form memo describe problem & action

"Intra-office memo" outline condition and resolution with attached copy of drwg. spec. or sketch

"Feedback constr. dept. description and resolution related to issued change orders

Evaluate

Director of arch/eng. will evaluate data provided in the three forms

Inform

Director of arch/eng. will provide data to operating staff to preclude recurrence in future projects

Action

Arch/site design group

Eng. design group

Arch/site prod. group

Eng. Prod. group

Field group

Figure 3-23 Feedback Program Diagram (Perkins & Will, Inc., Chicago, Illinois).

3-9.5 Furnishes valuable performance and construction information for young design professionals becoming involved in construction detailing.

Information gathered during field feedback will assist young designers in making critical decisions during detail development. Construction problems can be analyzed for suggestions that will guide development of new building details. The historic data bank will serve as an in-house continuing education base for the entire design team.

3-9.6 Facilitates in-house education programs for improving construction details.

Construction feedback can provide designers with a data base on material performance problems. Recurring problems logged in a historical file will help the designer pinpoint required detail improvement. Data recorded on construction site problems will help design professionals become more proficient in structuring the design process. As the data bank grows, it will be easier to conduct in-house education programs to improve with background knowledge the design of construction details. Such continuing education programs are ultimately most cost effective in that they lead to better quality construction documents and overall performance of the involved design department.

3-9.7 Supplies valuable information for developing an effective resources library.

A successful field evaluation plan and program will supply valuable data for expanding an in-house storage and retrieval program. A structured format for gathering field data can be linked to the in-house design library. Current data can be maintained for storage and retrieval if the system is planned, organized, and implemented. The growth of an information center can be limitless if

procedures are followed in accordance to an overall action plan.

The Field Evaluation Program should be structured to be interactive with and part of the Quality Control Program. Field feedback provides the base for developing quality control procedures. Both programs can benefit through a comprehensive approach to establishing communication lines for discovering design and construction problems. Figure 3-23 demonstrates one firm's approach to coordinating their feedback and quality control efforts.

3-10 DEVELOPING A DETAIL DATA CARD

A Detail Data Card should be developed along with each detail that is entered into the banking system. The purpose of a Detail Data Card is to document critical information recorded during detail development, construction, and evaluation. Each data card serves as a permanent record of how the detail has fulfilled its performance requirements. This information will provide the production staff with an up-to-date reference source for selecting acceptable details on future projects.

The Detail Data Card provides a written record of evaluation and detail approval for future reuse. Each card should be filed under appropriate retrieval descriptors selected for banking with detail. A descriptor selection will be made from the thesaurus of construction terms generated for the detail banking system. The Detail Data Card shown in Figure 3-24 identifies the critical information required for effective detail banking.

> A Detail Data Card is recommended for recording critical historic information on the status of a detail. However, this card can be eliminated from the process and retrieval descriptors placed on detail if user does not want to record and store historic data.

DETAIL DATA CARD

Bank Detail No.:	
Detail Title/Description:	

Scale:	Interior	Exterior

Development Data

Developed By:	Date:
Approved By:	Date:
Revisions:	Date:

Material Approval

Material Evaluation Approval No.	Date:
Manufacturer's Certification No.	Date:
Tests/Research	Date:

Project Identification

Project No.	Date:
Project Name	Location:
Soils:	Climate:	Rainfall:
Project Detail No.	Sheet No.:

Field Evaluation Record

	Period	Not Acceptable	Poor	Good	Excellent	Acceptable
1	Construction					
2	Five Year					
3	Ten Year					

Performance Problems

1	
2	
3	

Reuse of Detail

Reuse Instructions:	

1	Project No.: Project Name:	Project Detail No.: Date:	Sheet No.:
2	Project No.: Project Name:	Project Detail No.: Date:	Sheet No.:

Retrieval Descriptors

1.	5.
2.	6.
3.	7.
4.	8.

Figure 3-24 Detail Data Card.

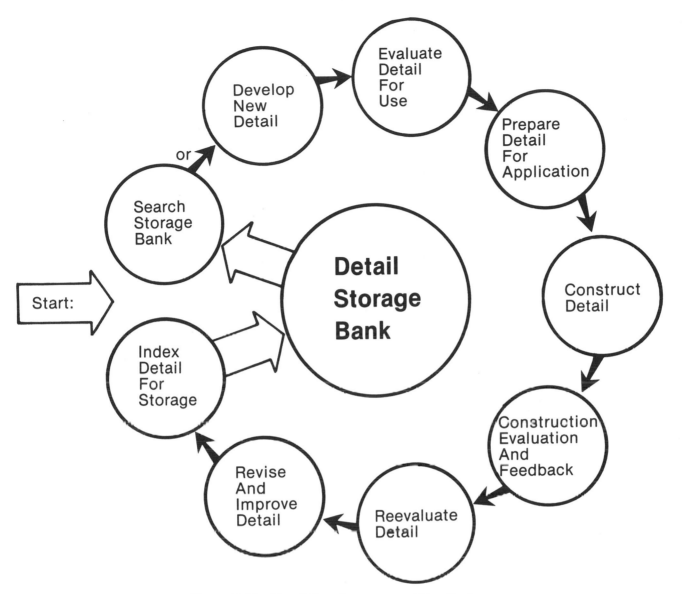

Figure 3-25   Detail Development and Use Cycle.

3-11   DETAIL DEVELOPMENT AND USE CYCLE

Once a detail has been applied, evaluated, and approved, it is ready for banking. However, the process of detail review and refinement goes on continuously for future applications. Every step from detail development to use should be systematically structured to reduce preparation errors and omissions. If the preceding guidelines are followed, the detail cycle shown in Figure 3-25 will emerge.

FOUR
DEVELOPMENT AND USE OF A CONSTRUCTION LANGUAGE

4-1 PURPOSE

To improve communication in the construction industry, it is necessary to develop and standardize a construction language. A formalized language will become more critical as more and more advanced communication technology becomes available for use in construction. Word processors, computers, and other types of automated communication equipment can greatly improve the effectiveness of information handling if a common language is generated in the construction industry.

Specialized data banks within the design professional's office can be made more efficient and effective if a planned communication language is generated. Studies of various organizations show that a core language is already in use within many disciplines of the building industry. A sampling of construction documents shows that as much as 75 percent of the terminology used in specifications and working drawings is common within many design offices.

Studies of legal problems in construction have shown that an increased number of claims have resulted from inconsistent and misinterpreted construction language. Many problems have resulted from professionals trying to invent new terms or use terminology that have multiple meanings. Inexperienced design professionals, who are free to generate their own terminology, have also added to the language confusion. In many offices design teams have a difficult time communicating because they lack a common language for sharing ideas. Specifications writers, consultants, and individuals responsible for pro-

ducing drawings are free to choose and select terms that they feel are best to express their individual ideas. As building technology and facility design become more complex, it becomes even more critical to coordinate ideas and express them through a common terminology.

An analysis of the current state-of-the-art on information handling shows that problems in language and lack of coordination have reduced the effective use of many existing systems. Most systems have generated unique languages, numbering systems, and other forms of abbreviation necessary to fulfill individual goals rather than system needs. By generating a unique indexing and retrieval system, they have forced users to develop specialized training programs to understand the information-handling process. As the complexity of these systems increases, their effective use has been found to decrease substantially. An improvement in system design is necessary to maximize the use of information generated in the construction field. Design professionals contemplating development and use of information-handling programs must make a coordinated effort to generate a common construction language that can serve many disciplines in construction industry.

Design professionals who depend on information transfer through details will benefit from the organization of an information-handling program that can manage reusable information. The present rate of information generated in design offices necessitates better communication to exchange data and save reinventing the wheel each time a new problem must be solved. Time, cost, and labor availability will no longer allow us to solve each problem in a unique manner just for the sake of being different. Effective problem solving in the future will depend on:

1. Researching historic data quickly and efficiently.
2. Analyzing solutions to similar problems.
3. Combining research information with user requirements to solve current problems.

To develop design solutions utilizing past experience, design offices will require the use of detail banking systems. Among design disciplines the exchange of ideas will be further enhanced by structuring a common construction language throughout the building industry.

4-2 PROBLEMS IN HANDLING INFORMATION

Research studies of information-handling programs within the construction industry have shown that a lack of coordination and standardization between indexing and retrieval systems have created major difficulties for users. Companies and organizations developing information-handling systems have chosen to create their own unique method of storing and retrieving information. These diverse approaches to information handling tend to confuse potential users because they require special training to understand the intent and purpose of each system. The obvious missing element in most systems is the common language or a standard vocabulary for communication. To this day we have not been able to coordinate a common language within construction industry that would help unite the many disciplines working on building construction. At the present time, contracts, working drawings, and other communication documents have been made more complex because of the inconsistent use of terminology.

With the lack of a common language, many design professionals have chosen to create their own storage and retrieval systems by generating specialized indexing systems. Uniqueness of these individualistic indexing systems has limited user understanding and reduced communication ef-

forts within the construction industry. These systems have required more training and individual attention because they are unique and not easily understood by different segments of the construction industry. The scale of our building industry is growing so rapidly that we can no longer afford to create individual systems which are only understood by a limited number of users. Planning ahead for the year 2000 will require greater cooperation to link common elements within information-handling programs so that larger-scale data centers can emerge.

Barriers to information storage and retrieval must be understood before developing new programs for information handling. Some of the common barriers that have been identified through various studies are projected as follows:

1. Inability to communicate because of a lack of standard terminology or a common language.
2. Information classification and indexing systems are individually designed and unrelated to any common means of communication.
3. Many different retrieval systems exist within design offices nationwide.
4. Information is not properly prepared for storage and retrieval. There is a lack of standardization and uniform format for storage.
5. Information is not able to be cross-referenced within the system, therefore making it difficult to retrieve related types of information.
6. Information-handling system design has been limited to the goals of individuals, sources of funding, and preconceived ideas of users' needs.
7. Designers of information-handling systems have tended to work in a vacuum without attempting to incorporate past experience from information-handling problems.

These are some of the barriers that keep appearing when information-handling problems are discussed and identified by design professionals. By analyzing past problems and identifying user needs, future system designers will be able to structure information-handling systems that can be easily understood on a national scale.

Information-handling systems must consider the information search process before attempting to develop an indexing system. When the human mind translates an idea from a picture, it turns out to be the words for creating a construction language. This language formulates the verbal description that we use to coordinate our construction documents. If the language or terminology is not clear or precise, confusion will develop. This type of confusion has also been introduced into information-handling systems through use of symbolism, synonyms, abbreviations, and acronyms that tend to confuse users and searchers of information. It has been found that with a well-structured vocabulary the user can project from an idea to a word association and finally into a thesaurus of acceptable terminology that can be used to retrieve a particular item of information. When this train of thought is broken with symbolism or some other indirect means of association, it will cause major confusion and will tend to delay and thought process necessary to achieve an effective and efficient search. The thought process from idea to search is diagramed in Figure 4-1.

Future information-handling systems must streamline the search process to relate to the user's description of the problem. Simple, direct, and straightforward indexing can greatly enhance turnaround time for information input and output in detail banking.

Analysis of existing information-handling systems have identified several major problem areas that must be considered when structuring a national or in-house construction language [8]. A closer look at the common barriers can provide us with directives for structuring a language that will be acceptable for use in storing design information. The following considerations highlight the critical issues that need to be worked out in system design.

4-2.1 Existing inhouse systems have limited flexibility and are generally restrictive.

Existing information systems have limited flexibility to expand beyond a predetermined subject area or classification of information. Most systems are restricted by the codification and classification of a narrow field of information. Systems restricted to one area of information are forced to add subsystems when a new area of information is created. As subsystems are added, greater confusion is created for the user. It is important to keep in mind that a system be designed so that it is open-ended and flexible to receive a variety of information types over a long period of time.

4-2.2 Inability to cross-reference information limits system use.

Many existing systems operate in only one direction. Information is entered into one category and can only be retrieved through the use of a specific coding for that category. One-direction retrieval causes valuable related items of information to be lost or unavailable when required by the design professional. It is important that information identified in one category be cross-referenced into several other categories that have a relationship to the main category under which the information was indexed. Terminology and language should be structured to provide several levels of entry into one category of information. The success and efficiency of a system is dependent on the ability of the user to reach an area of information through several directions of entry.

4-2.3 The absence of a standardized terminology in the construction industry has resulted in inconsistent methods of indexing information.

Design professionals creating information-handling systems within offices have lacked the guidelines and vocabulary to work effectively with different types of construction information. As a result they have resorted to complex numbering systems and alphanumeric symbols which help to complicate the indexing system. This approach to indexing increases time and leads to indecision on the part of the indexer in determining which

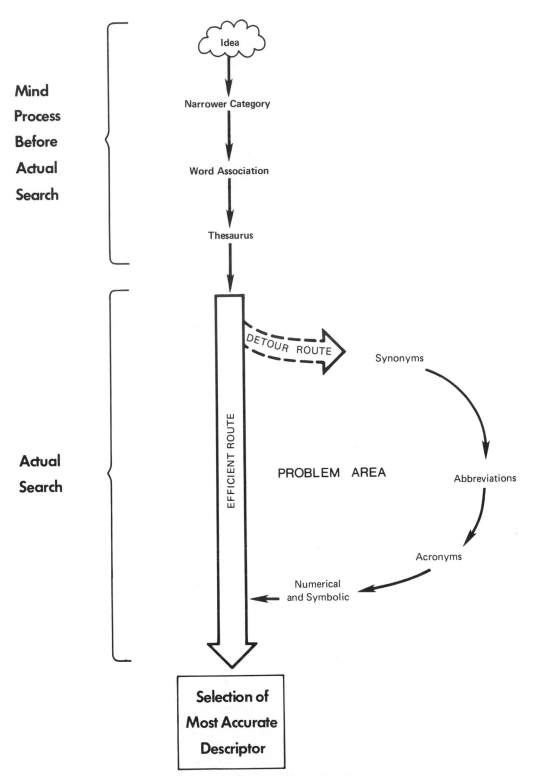

Figure 4-1 Information Search Process. (Philip M. Bennett and Judy A. Jones. *Construction Information Systems Study.* The University of Wisconsin—Extension. Department of Engineering and Applied Science. For the Construction Sciences Research Foundation, Washington, D.C. © 1978.)

category or numbering system should be used for a specific area of information. In some cases information ends up being misplaced because the document was improperly indexed when the numbering system did not allow for a consistent indexing. In programs where key words or selected terminology were used, they were not representative of a common language generated within the construction industry. A common language with selected terminology must be evaluated and selected on the basis of acceptance from within different segments of the construction industry.

4-2.4 Symbolism creates confusion for users of information systems.

Existing systems have utilized letters, numerals, and acronyms to designate different subject areas of information. These abstract forms of indexing cause users to spend larger amounts of time becoming familiar with each system and subsystem. Individualized systems that consist of word groupings associated with distinct and special areas of study can limit the flexibility of indexing information. Users of these systems must become totally familiar with coding, word groupings, and other numerical methods before the information can be selected for retrieval. As a general rule symbolism confuses the information searcher and makes it more troublesome to understand each unique system and therefore reduces the level of interest expressed by a user group. Every effort must be made to reduce symbolism in indexing information for storage and retrieval.

4-2.5 Acryonyms and abbreviations cause confusion in indexing information.

Increased use of acronyms and abbreviations in the design field can cause confusion in indexing information. Once a large number of acronyms are generated in a particular area of building construction, they become misleading and dis-

courage efficient use of an information-handling program. The overuse of acronyms and abbreviations causes a greater amount of time lost determining the actual meaning associated with each term. As a result confusion, misunderstanding, and misinterpretation of critical information occurs in both the storage and retrieval activities. Overuse of abbreviations and acronyms cause systems to be only understood by the individuals who developed their programs.

4-2.6 Synonyms complicate an information search process.

A synonym, though an alternate choice of a word, often creates confusion in determining the intended meaning or proper usage of a term. Synonyms must be carefully selected and screened for proper use in indexing information. It is important to identify the intended definition and the proper use of the term selected for the information-handling system. The selector of terminology must structure a method for handling synonyms in accordance with the guidelines for generating a construction language [9].

4-2.7 Indexing with specific categories of information that overlap each other can create confusion on the part of the indexer.

Systems that have developed a specific number of categories for storing information can be restrictive to both the information indexer and the searcher. If these categories overlap each other, the indexer must question which category should be selected for the information. If the choice involves two or three categories, it will greatly confuse the indexing process. At some point a decision must be made as to which category or categories will be selected for storing information. Unless the user of the information is aware of the indexer's guidelines for selecting the category, it will be virtually impossible for the searcher easily to reach the information under search. Multicategory indexing can create confusion, misunderstandings, and inaccessible information if directions are unclear. Present systems that are currently being used with specific categories have already created confusion on the part of the indexer and the retriever of information. Users of these systems have indicated hesitation as to which category should be selected for storing information. *Future systems must eliminate this problem by creating greater flexibility in indexing information and not restricting information to a particular category.* This type of problem points the way toward the creation of a construction language that allows a particular item of information to be indexed in accordance with the terminology that describes its subject areas.

4-3 SUPPORT FOR A CONSTRUCTION LANGUAGE

Word processors and computers are being designed to work effectively with a vocabulary and a construction language. By developing a construction language, design professionals can create a system for interdisciplinary communications to take place in the construction industry. Selected terminology that is acceptable to all design and construction disciplines will provide a basis for information exchange and will allow computer programs to be prepared in a format that standardizes information retrieval.

Studies of foreign information systems have shown that standardized vocabularies and descriptor terms from construction industry thesauri are much more effective in gaining access to construction information than conventional indexing systems [8]. Key words and descriptor terms indentified for each item of information make indexing and retrieval easier for the user. Development of a common language is the first essential step in standardization of in-house information-handling programs.

The ultimate goal is to encourage development of an acceptable universal construction language. CSI, AIA, and other construction-related associations should be organizing a coordinated language on a national level. Since efforts to date have only produced fragmented results, we must encourage organizations to take steps in using existing resources, such as CSI's Uniform Construction Index and the Canadian Construction Thesaurus, to establish a workable language [10], [11].

Motivating design professionals to work together in building a construction language is essential to improving communication. Since a national construction languge is not being developed, it is incumbent on local or regional organizations to take the initiative to begin structuring a common language among themselves. Coordination between individual firms and design associations will help to develop a universal language that will enable information to be processed by using a consistent terminology to ensure standard information indexing and retrieval.

A 1978 study of characteristics common to 14 U.S. information systems in different disciplines has highlighted the successful use of a standard terminology for indexing [8]. Many of these systems generated a subject area thesaurus or key word system to aid in the storage and retrieval of information. Six of the systems studied indicated no major problems in handling information on a subject area basis. On the other hand, one system using random indexing identified several problems. *The remaining 7 of the 14 systems indicated the need to standardize further their vocabulary and control it by use of a thesaurus.*

In 1967 to 1970 I was involved in coordinating the development of the construction-related terms identified in the ERIC Thesaurus of Descriptors for the Clearing House on Educational Facilities. This thesaurus contained the language necessary for storing and retrieving educational facilities design information used nationwide. Under this program a manual and computer system was designed for document storage and retrieval. To facilitate information indexing, a standardized terminology was created by selecting acceptable terms identified within the literature. Common and standardized terminology was identified in each document abstracted for inclusion into the ERIC system. Each term selected was structured into a family of related terms and evaluated according to rules of the Joint Council of Engineers for structuring a thesaurus [12]. An example of this thesaurus is shown in Figure 4-2. The main thesaurus of educational facility design terminology was utilized for indexing and retrieving approximately 4000 documents [13]. This system worked effectively on both a manual and computer-based retrieval program.

Why is a construction language the key to effective information handling in the design profession? Present-day construction terminology:

1. Provides the one element that links all segments of the construction industry and serves as a communication pipeline for information exchange. Figure 4-3 shows a construction language as the project life line.

2. Establishes the means for accomplishing a common understanding in the transfer of building technology.

3. Creates a picture and word association that forms the simplest means for conveying an idea or concept in the design profession.

4. Provides a means to control indexing of information for more effective storage and retrieval.

5. Develops the basis for small and large organizations to process in-house information in a controlled manner.

6. Encourages the development of a standardized terminology for construction information-handling programs.

7. Establishes a consistent language for developing uniformity in communication through working drawings and specifications.

Additional support criteria can be identified for creating a common construction language such as to *(1) provide a means to control indexing of information for more effective storage and retrieval and (2) encourage the development of a standardized terminology for construction information handling programs. However, the most important reason is identified in item 7, the wide use of the terminology associated with construction documents.* Efforts to develop a standardized terminology are displayed in Figure 4-4.

An examination of the working drawings and specifications can provide us with more specific reasons for creating a construction language:

1. Uniformity in producing working drawings.

   a. All project team members can work with the same vocabulary and terminology.

   b. The language supports effective coordination of working drawings and specifications.

   c. A greater understanding of requirements and materials can be achieved by the contractors and construction workers.

   d. Standardized terms will help to reduce misunderstandings and confusion in working with the drawings and specifications.

THESAURUS OF ERIC DESCRIPTORS

APPLIED READING	440				RT	ABILITY
BT	READING					ACHIEVEMENT
APPLIED RESEARCH						APTITUDE TESTS
	USE RESEARCH					ASPIRATION
APPRENTICE EDUCATION						COGNITIVE ABILITY
	USE APPRENTICESHIPS					EXPECTATION
APPRENTICE PROGRAMS						PERFORMANCE
	USE APPRENTICESHIPS					STUDENTS
APPRENTICESHIPS	270					TALENT
UF	APPRENTICE EDUCATION					
	APPRENTICE PROGRAMS					
	APPRENTICESHIP TRAINING					

NT	RESEARCH APPRENTICESHIPS
RT	INDUSTRIAL TRAINING
	INPLANT PROGRAMS
	ON THE JOB TRAINING
	TRADE AND INDUSTRIAL
	EDUCATION
	TRAINEES
	VOCATIONAL EDUCATION
	WORK EXPERIENCE PROGRAMS
APPRENTICESHIP TRAINING	
	USE APPRENTICESHIPS
APTITUDE	010
NT	ACADEMIC APTITUDE
	VOCATIONAL APTITUDE

APTITUDE TESTS	520
BT	TESTS
RT	APTITUDE
	OCCUPATIONAL TESTS

Figure 4-2 Example of the Thesaurus of ERIC Descriptors.

ARABIC 300
BT SEMITIC LANGUAGES
RT ARABS

ARABS 380
RT ARABIC
ETHNIC GROUPS
NON WESTERN CIVILIZATION
RACIAL CHARACTERISTICS

ARBITRATION 150
RT BOARD OF EDUCATION POLICY
COLLECTIVE BARGAINING
COLLECTIVE NEGOTIATION
EMPLOYMENT PROBLEMS
GRIEVANCE PROCEDURES
LABOR DEMANDS
LABOR ECONOMICS
LABOR LEGISLATION
LABOR PROBLEMS
NEGOTIATION AGREEMENTS
NEGOTIATION IMPASSES
SANCTIONS
STRIKES
TEACHER ASSOCIATIONS
TEACHER MILITANCY
TEACHER STRIKES
UNIONS

ARCHAEOLOGY 480
BT SOCIAL SCIENCES
RT ANCIENT HISTORY
ANTHROPOLOGY
ETHNOLOGY
PALEONTOLOGY

▷ ARCHITECTS 380
BT PROFESSIONAL PERSONNEL
RT ARCHITECTURAL EDUCATION
ARCHITECTURE
BUILDING DESIGN
CONSTRUCTION INDUSTRY
PROFESSIONAL OCCUPATIONS

▷ ARCHITECTURAL BARRIERS 210
SN BUILDING ELEMENTS WHICH
BECOME OBSTACLES TO
PHYSICALLY HANDICAPPED
PERSONS
RT ARCHITECTURAL ELEMENTS
BUILDING DESIGN

▷ ARCHITECTURAL CHANGES
USE BUILDING DESIGN

▷ ARCHITECTURAL CHARACTER 210
SN STYLISTIC EXPRESSION OF
VERTICALITY, SCALE, RICHNESS,
VARIETY, AND UNITY INHERENT
TO ARCHITECTURAL
TRADITION
UF ARCHITECTURAL STYLE

▷ ARCHITECTURAL TRADITION
RT ARCHITECTURAL ELEMENTS
ARCHITECTURE
BUILDING DESIGN
CAMPUS PLANNING
DESIGN PREFERENCES

▷ ARCHITECTURAL DESIGN
USE BUILDING DESIGN

▷ ARCHITECTURAL DRAFTING 350
BT DRAFTING
RT ARCHITECTURAL EDUCATION
ARCHITECTURAL PROGRAMING
ARCHITECTURE
BUILDING PLANS

▷ ARCHITECTURAL EDUCATION 140
BT PROFESSIONAL EDUCATION
RT ARCHITECTS
ARCHITECTURAL DRAFTING
ARCHITECTURAL RESEARCH
ARCHITECTURE
ART EDUCATION
BUILDING DESIGN
TECHNICAL EDUCATION

▷ ARCHITECTURAL ELEMENTS 210
SN BUILDING MATERIALS THAT
SATISFY THE ARCHITECTURAL
REQUIREMENTS OF BUILDING
CONSTRUCTION
NT DOORS
RT ARCHITECTURAL BARRIERS
ARCHITECTURAL CHARACTER
ARCHITECTURE
BUILDING DESIGN
BUILDING MATERIALS
BUILDINGS
CONSTRUCTION (PROCESS)
CONSTRUCTION NEEDS
MASONRY
PREFABRICATION
PRESTRESSED CONCRETE
SCHOOL BUILDINGS
SCHOOL CONSTRUCTION
SCHOOL DESIGN

▷ ARCHITECTURAL PROGRAMING 210
SN THE PROCESS OF
IDENTIFICATION AND
SYSTEMATIC ORGANIZATION
OF THE FUNCTIONAL,
ARCHITECTURAL, STRUC-
TURAL, MECHANICAL, AND
ESTHETIC CRITERIA WHICH
INFLUENCE DECISION MAKING
FOR THE DESIGN OF A
FUNCTIONAL SPACE,
BUILDING, OR FACILITY

BT PROGRAMING
RT ARCHITECTURAL DRAFTING
ARCHITECTURAL RESEARCH
ARCHITECTURE
DECISION MAKING
DESIGN NEEDS
SYSTEMS ANALYSIS

▷ ARCHITECTURAL RESEARCH 450
BT RESEARCH
RT ARCHITECTURAL EDUCATION
ARCHITECTURAL PROGRAMING
BEHAVIORAL SCIENCE RESEARCH
BUILDING MATERIALS
COMPONENT BUILDING SYSTEMS
COMPUTERS
DESIGN
ENVIRONMENTAL INFLUENCES
ENVIRONMENTAL RESEARCH
MECHANICAL EQUIPMENT
PHYSICAL ENVIRONMENT
STRUCTURAL BUILDING SYSTEMS

▷ ARCHITECTURAL STYLE
USE ARCHITECTURAL CHARACTER

▷ ARCHITECTURAL TRADITION
USE ARCHITECTURAL CHARACTER

▷ ARCHITECTURE 210
NT SCHOOL ARCHITECTURE
RT ACOUSTICS
ARCHITECTS
ARCHITECTURAL CHARACTER
ARCHITECTURAL DRAFTING
ARCHITECTURAL EDUCATION
ARCHITECTURAL ELEMENTS
ARCHITECTURAL PROGRAMING
ART EDUCATION
BUILDING DESIGN
COMPONENT BUILDING SYSTEMS
DESIGN
DESIGN NEEDS
DESIGN PREFERENCES
FINE ARTS
INTERIOR DESIGN
LIGHTING DESIGN
SPATIAL RELATIONSHIP
STRUCTURAL BUILDING SYSTEMS

▷ ARCHIVES 210
RT COLLEGE LIBRARIES
GOVERNMENT LIBRARIES
LAW LIBRARIES
LIBRARIES
LIBRARY COLLECTIONS
MEDICAL LIBRARIES
NATIONAL LIBRARIES
PUBLIC LIBRARIES
RECORDS (FORMS)
STATE LIBRARIES

Note the potential to structure families of related descriptor
terms. This capability provides for cross-referencing and multi-
access to information.
▷ Architectural related terms.

Figure 4-2 (Continued)

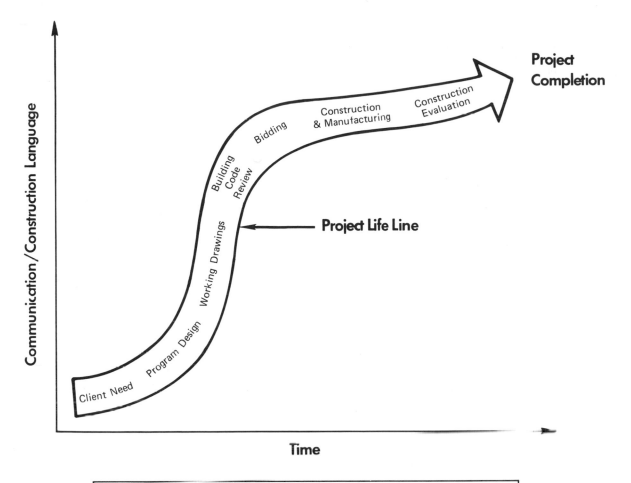

A Construction Language is the Life Line of the Project Development Process

Figure 4-3 Importance of Establishing A Construction Language. (Philip M. Bennett and Judy A. Jones. *Construction Information Systems Study*. The University of Wisconsin—Extension. Department of Engineering and Applied Science. For the Construction Sciences Research Foundation, Washington, D.C. © 1978.)

2. Storage and retrieval of construction information.

   a. Construction information indexing is accomplished by using acceptable terminology within the construction industry.

   b. A uniform language provides for easy indexing and entry into data banks.

   c. Different disciplines within the construction industry can work with accepted terminology.

   d. All entries into the data bank are made with accepted descriptor terms.

The greatest area of support for a construction language can be derived from the physical and behavioral actions of the human body. Human memory and brain functions establish the basis of information processing. They dictate the range of communication possible within the human system. A further analysis of the human mind identifies the natural process of a search as it

TERMINOLOGY

"Don't misspell — don't misuse . . .

A construction drawing not only conveys information by lines, symbols, dimensions and graphics, but in substantial part by lettered notations. These notations must be of consistent and clear nomenclature. It is not intended that they become literary milestones. Terminology must also be consistent between agreements, drawings, specifications and cost estimates. The use of several words for one meaning can only lead to different interpretations by the many parties who use the documents.

The list of terms is not intended to be a dictionary.

RECOMMENDATIONS

The recommended uses of words shown in the following list have been limited to those which experience indicates are most often misused and sources of confusion. Words which are consistently properly used and understood are not included.

Specialized projects may require that particular uses be established for other words in order to keep notations consistent throughout the documents.

Capitalized terms denote recommended usage. Lower case terms denote undesirable or misleading variants which should be avoided in drawings and specifications. Comments are for assistance in defining terms more clearly.

ACCESS DOOR	Use only for small doors not included in door schedules. Prefabricated assembly including frame and door.
ACCESS FLOOR	Not "pedestal floor", "free access floor", or "computer floor".
ACCESS PANEL	A section of finish which can be opened.
ACOUSTICAL PANEL ACOUSTICAL TILE	Not "acoustic", or "acoustic board".
ACOUSTICAL SEALER	Non-hardening calking or sponge tape used to se partitions to structural ceiling, walls and floor to reduce sound transmission. Not "sound calking"
acoustical underlayment	Use **SOUND DEADENING BOARD**.
ADHESIVE	Do not use "glue" unless the particular adhesiv is commonly called "glue" (such as for wood). Do not use "cement", "paste" or "mastic".
AGGREGATE BASE	Not "crushed rock", "gravel", "rock sub-base"
ALTER ALTERATION	Not "remodel".
ALTERNATE	Use for pricing. Use **OPTION** when there is a choice.
ANCHOR BOLT	A bolt that is embedded in masonry or cast-in place in concrete. Do not use to describe an expansion bolt or a bolt in an expansion shiel
ANODIZE	Not "alumilite", "alodize", "kalcolor" or "duranodic", etc. Avoid use on drawings.
apply	Use **INSTALL**. Avoid use on drawings.
asbestos cement	Use **MINERAL FIBER**.
AS-BUILT DRAWING	A drawing or print marked by the Contracto show actual conditions as constructed. For Architect's drawing, see **RECORD DRAWINC**
ASPHALT CONCRETE	Not "bituminous concrete" or "asphalt pavin
asphalt roofing	Use **BUILT-UP ROOFING**.

COMMITTEE ON PRODUCTION OFFICE PROCEDURES

©**JULY 1980**
NORTHERN CALIFORNIA CHAPTER
AMERICAN INSTITUTE OF ARCHITEC

Figure 4-4 Development of Standardized Terminology. (Northern California Chapter of the American Institute of Architects. San Francisco, California.)

begins with a basic idea or premise and is further translated into words. These words on a broader scale form the common language used for communication within a given discipline. Assuming a language is understood by the researcher, it is then possible to select the proper term or phrase associated with an actual information search. The actual information search process is diagramed in Figure 4-5.

4-4   PROCEDURES FOR SELECTING AND DEVELOPING A NATIONAL OR IN-HOUSE LANGUAGE FOR DETAIL BANKING

A construction language should be made up of terminology that is recognized and accepted as part of the English language. Terms should be selected that are commonly accepted and used by each discipline within the construction industry. The vocabulary should represent specific tasks and activities that are carried out by the prime user groups represented within the construction field. This encompasses design, manufacturing, construction and building codes. All terms generated must be evaluated and proved acceptable on a large geographical base. The construction language and thesaurus should be structured by incorporating the terminology standards established by AIA, CSI, and other construction-related associations. *However, we need additional freedom to go beyond the limits of established systems to achieve total flexibility for indexing graphic information.*

Creating an acceptable national or in-house construction language should begin by selecting an advisory committee made up of different disciplines within the design office. This committee should be led by a director who can manage the research analysis and development process. Minimum time and effort will be required to structure a common language, if a plan of action is based on goals and objectives established for the detail banking system. *One trained individual can become very efficient at indexing details for the banking system.*

Two major stages of activity must be undertaken.

1. Research, evaluate and identify acceptable terms.
2. Develop a committee review process to establish acceptability for each term used in detail banking.

It is important that the committee and its director acquire current documentation to support terms selected for indexing and creating a common language. To document support for a descriptor, it is recommended that the approval form in Figure 4-6 be prepared.

The form in Figure 4-6 is designed to record support for an acceptable term for detail banking. Not all terms would require extensive research or verification for use in the construction field. Many construction terms are already generally accepted in the construction industry and therefore require no further investigation. This form is designed to record the research and document the approval of contradictory terms and synonyms, those commonly misused by many individuals working in the design profession.

Specific steps and procedures must be followed in order to create a construction language that is acceptable and functional in the design office. The following procedures should be considered as guidelines to accomplishing the goal.

4-4.1   Review specifications and working drawings for common and repetitive terminology.

Assign personnel to collect and research past projects for common terminology. The historic files of specifications and working drawings will provide an excellent base for demonstrating the repetitive use of terms. It is important in this step to identify standard terms used frequently throughout the documents. The greater frequency of use will demonstrate the importance of a term and also provide the basis for establishing it as a standard for future use.

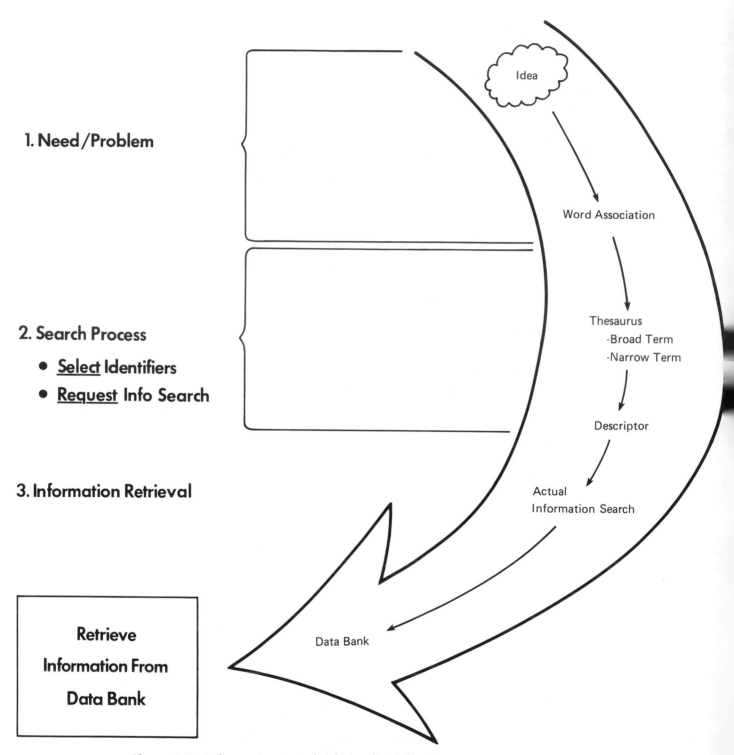

1. Need/Problem

2. Search Process
- **Select** Identifiers
- **Request** Info Search

3. Information Retrieval

Retrieve
Information From
Data Bank

Idea

Word Association

Thesaurus
-Broad Term
-Narrow Term

Descriptor

Actual
Information Search

Data Bank

Figure 4-5 Information Search Channel. (Philip M. Bennett and Judy A. Jones. *Construction Information Systems Study.* The University of Wisconsin—Extension. Department of Engineering and Applied Science. For the Construction Sciences Research Foundation, Washington, D.C. © 1978.)

CONSTRUCTION LANGUAGE APPROVAL FORM

Present Use of Term

Descriptor Term _____
Variations of Term _____
Definition of Term _____

Approved Use of Term

Descriptor Term: _____
Approved Definition: _____
_____
Approval Date: _____

Research Data

Research By _____
Date: _____

Reference Sources: Accepted Term

Dictionaries   _____   _____
Thesauri   _____   _____
Uniform Construction
Index   _____   _____
National Organizations _____   _____
Federal Specifications _____   _____
Building Codes   _____   _____
ASTM and ANSI   _____   _____
Manufacturers   _____   _____
Publications   _____   _____
Special Authorities   _____   _____
Other User Groups   _____   _____

Figure 4-6   Construction Language Approval Form.

4-4.2 Research standard construction terminology generated by different agencies, associations, and government programs.

A search of existing standards and specifications already accepted by the construction industry can provide the basis for creating a common language. Many building research centers are developing documents and utilizing common terminology to present innovative techniques in construction. Terms that have common usage can provide an excellent base for developing an in-house program. Some of the specific aids for developing a common construction language are identified as follows:

▷ Technical research documents generated by building research stations around the world.
▷ National standards and building codes.
▷ English language dictionaries and thesauri.
▷ Building construction thesauri (these provide the most direct source for acceptable terms to be used within the design office).

The terms identified in these documents can be evaluated for in-house detail banking. Several of the following thesauri should be considered valuable in structuring a construction language:

▷ The *Uniform Construction Index.*
▷ The *Canadian Thesaurus of Construction Science and Technology.*
▷ The English *Construction Industry Thesaurus.*
▷ The ERIC *Thesaurus of ERIC Descriptors.*
▷ The *Thesaurus of Engineering and Scientific Terminology.*

These and other technical thesauri should be utilized in generating a standard in-house construction terminology [10], [11], [14], [15], [16]. Cover pages identifying appropriate thesauri are shown in Figure 4-7.

4-4.3 Select terms that have frequent recurrence in construction documents.

To begin a construction language, it is important to select those terms that have the most frequent usage within the construction documents. The terms selected should have a direct use for those items of information that are requiring indexing and storage. As the system grows, terminology can be added as required to index new information generated in working drawing production.

4-4.4 Select terms that are appropriate and acceptable subject areas in the design office.

All terms selected for potential reuse should be evaluated in accordance to the thesaurus development guidelines. In the first level of selection, it is important to select the generic terms or broadest concepts identified within the construction document. As the system expands, more attention should be given to narrower terms and specific subject areas that develop the basis for cross-referencing information within the system. Greater support for selecting a term will increase its chances of being nationally accepted in future construction language programs.

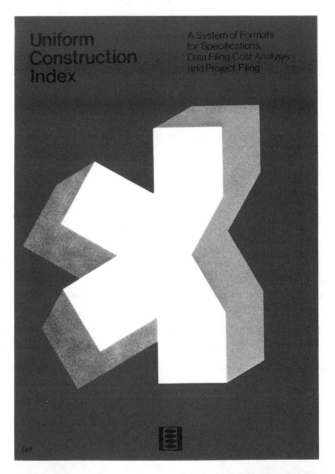

Figure 4-7 Cover Pages of Existing Thesauri Appropriate for Use in A Construction Language. (a) The *Uniform Construction Index.* (b) The *Canadian Construction Thesaurus.* (c) The *English Construction Industry Thesaurus.* (d) The *ERIC Thesaurus.*

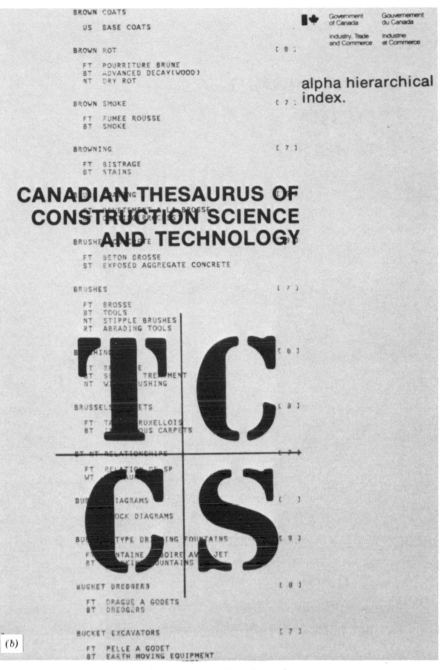

Figure 4-7 *(Continued)*

4-4.5 Establish a terminology evaluation process for new terms.

A standard format for selecting terms should be developed by the control committee. Procedures should be extracted from guidelines prepared in the field of library science. It is important to develop vocabulary control procedures for selecting terms that are currently acceptable within a discipline area. The acceptance of terms should be based on tests, research, accountability, history of use, and a general understanding of the specific

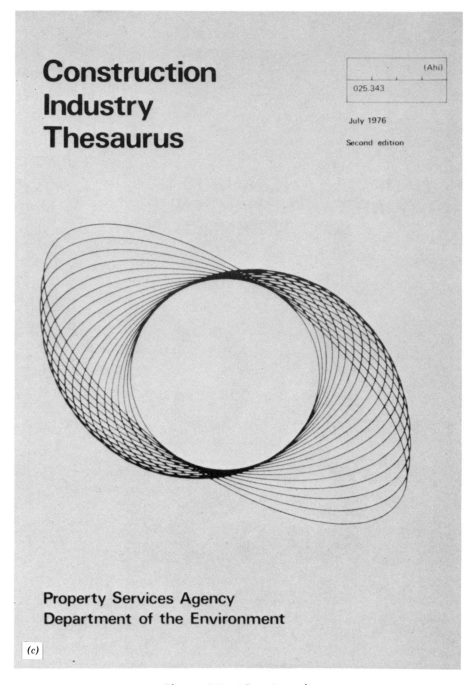

Figure 4-7   (Continued)

terminology. Sufficient guidelines exist within the information-handling discipline to create standards that are acceptable for a national or in-house program.

4-4.6   Develop definitions for selected terms.

Terms used in the construction language should be defined for consistency in future use. To create

THESAURUS OF ERIC DESCRIPTORS

WORKING COPY

DESCRIPTOR LISTING

June 1970

This is a total cumulative edition of the Thesaurus of ERIC Descriptors which supersedes previous issues and includes new descriptors introduced in indexing documents for the issue of Research in Education (RIE) carrying the data above. It is being provided as an inhouse working copy.

A rotated Descriptor Display is not provided with this issue. A May 1969 cumulative Rotated Descriptor Display was distributed to each ERIC Clearinghouse and should be retained.

ERIC PROCESSING AND REFERENCE FACILITY

(d) RATED FOR U.S. OFFICE OF EDUCATION BY LEASCO SYSTEMS & RESEARCH CORPORATION
4833 RUGBY AVENUE, BETHESDA, MARYLAND 20014

Figure 4-7 (Continued)

a common understanding of the language, it is also necessary to define difficult terms in a controlled standard format. Definitions will reduce confusion and misinterpretations, while providing for consistent and acceptable term usage in the construction industry. Developing a manual of definitions will aid users in answering language questions.

The *Glossary of Construction Specifications Terminology* prepared by the District of Columbia Metropolitan Chapter of the Construction Spec-

ifications Institute, Washington, D.C., is an excellent example of a major effort to reduce misinterpretations of construction terms [17]. Each Division and Section has commonly used terms defined. The example definitions demonstrate control achieved for term usage:

Division 4. REGLET. A narrow, flat recessed molding of rectangular profile (MIA). Section 04422 **MARBLE**.

Division 6. BRIDGING. Short wood or metal braces or struts crosswise between joists to hold them in line. Bridging may be solid, or may be crossed struts (APA). Section 06100 **ROUGH CARPENTRY**.

Division 9. CONTROL JOINT. A predetermined opening between adjoining panels or surfaces at regular intervals to relieve the stresses of expansion and contraction transverse to the joint in large wall and ceiling area (GA). Section 09250 **GYPSUM WALLBOARD**.

Another excellent example of construction term definitions can be found in the British *Construction Industry Thesaurus—Definitions of Construction Terms* [18].

4-4.7 Group-related families of terminology to cross-reference information in the detail banking system.

The key to cross-referencing information is to create common families of terminology that will guide researchers to the appropriate items of information. An effective cross-referencing system includes descriptor term relationships. In the system all terms—broad, narrow, as well as related—can lead the research to the appropriate descriptors for retrieving a particular item of information. The example that follows shows a family relationship for the term "roofs."

A Family Relationship of Descriptor Terms: Developing Categories and Families of Descriptor Terms

▷ Broad term = Roofs.
▷ Narrow terms = Flat roofs,
　　　　　　　　　Pitched roofs,
　　　　　　　　　Domes.

4-4.8 Develop classifications of terminology if the system is going to be used for other areas of information handling.

Design organizations have a variety of activities, procedures, and data bases that require information handling and storage. If the firm is anticipating development of other information-handling areas, it is recommended that common terms used within the organization be classified into specialized areas of use. Through the development of a common language, a large-scale information-handling bank can be structured with the same guidelines and procedures developed for the detail banking system. Some important areas of terminology classification are identified as follows:

▷ Business management.
▷ Construction methods and technology.
▷ Contractual and legal.
▷ Product and material.
▷ Design information (*detail banking system*).
▷ Human factors and behavior.
▷ Codes and standards.
▷ Costs.
▷ Maintenance.

These are valuable design categories to be considered in developing a common language for information handling. A common language in these categories will provide the link to all information-handling areas within the design office. The terms used in these areas can also further the development of a standard language for indexing design-related information.

4-4.9 Organize a mini-thesaurus that will serve as an in-house office manual for specifications and working drawings.

Terminology selected and approved should be documented and defined in a construction thesaurus. Initially the thesaurus will be limited to only those terms that have recurring use within the design firm. As the detail system grows, new terms will be selected and evaluated for input to a larger construction thesaurus. Members of the design team will be required to use the established thesaurus for selecting appropriate terminology in developing construction documents. Consultants, designers, specifications writers, and draftpersons will also be required to use the construction thesaurus for vocabulary selection. An example of a mini-thesaurus is shown in Section 4-5.2.

4-4.10 Develop a manual of guidelines and procedures for preparing design information and selecting appropriate terminology.

Office manuals should contain guidelines and steps to prepare details and select appropriate vocabulary for consistency in detail banking. It is critical that information be prepared following a standard procedure and indexed with the subject area descriptors identified in a thesaurus of construction language. Information indexers should systematically select descriptors appropriate for each detail prepared for storage.

Developing a construction language appropriate for in-house detail banking must be considered an ongoing process that is standardized and controlled so that uniform and consistent practices emerge. The language developed should be open-ended so that the selection of new terms will be acceptable over a long period of time. One of the major advantages of a construction language is that it can reflect the continual change of technology, terminology, and innovation within the design profession. Most systems do not allow this level of flexibility and therefore are found to be inadequate after changes in technology take place. The language approach improves communication by automation and allows the most effective level of information handling to be established.

4-5 STRUCTURING A THESAURUS OF CONSTRUCTION TERMINOLOGY

A construction thesaurus should contain the terminology appropriate for detail banking. All terms selected, evaluated, and screened for potential use should be entered into it along with definitions. A thesaurus of terms will provide a framework for information indexing, storage, and retrieval as well as in-house preparation of working drawings and specifications. As the construction thesaurus grows, it will provide thousands of subject areas under which information can be stored and retrieved. In order to ensure maximum efficiency, a thesaurus should be structured to avoid abbreviations, symbols, acronyms, and synonyms.

One of the main purposes of a thesaurus is to list synonyms or nearly synonymous terms that can cause confusion and misinterpretation. This objective will be kept in mind while indexing and selecting information. Users of information-handling systems have indicated that the main problem in the search and retrieval process is the overabundance of alternate choices of synonyms. It has been found that the search process is made more difficult and the success rate is brought noticeably down if the use of synonyms is not controlled in the indexing process. The thesaurus can be developed to control the quality of vocabulary by eliminating troublesome synonyms through the use of families of terminology. All terms that are synonymous will be dropped and replaced with specific subject headings. The remaining terms will indicate which use of the term has been selected along with the intended definition to be used in future information handling.

A thesaurus attempts to control the language of a subject field. Vocabulary control is necessary due to the presence of ambiguous meanings in a language. A thesaurus does not define a term; rather, meanings are implied by associating one term with another related term. This system of information handling is dependent on a common language so that all users will be able to work with similar indexing and retrieval terminology. In this way the searcher only needs to know the subject area; the computer or librarian can do the rest.

A solution to the current problems in storage and retrieval systems is the "direct approach" in thesaurus usage. The use of a subject area descriptor is the best method of indexing terms. *Rules for Thesaurus Preparation*, published by the U.S. Department of Health, Education, and Welfare, offers certain guidelines for the process [9]. The following is a summary of these steps:

1. *Descriptors* should represent important concepts found in the literature rather than concepts derived independently. They should also reflect the language used in the literature to describe such concepts.
2. *Used For* (UF) indicates preferred usage.
3. *Broader Term* (BT) is a cross-reference term that indicates a more inclusive class relationship that may exist among descriptors.
4. *Narrower Term* (NT) is a cross-reference term that indicates a more restrictive class relationship that may exist among descriptors.
5. *Related Term* (RT) is a cross-reference term used as guide from a given descriptor to other descriptors that are closely related.
6. *Scope Notes* (SN) is a brief statement of the intended usage of a descriptor.

It is appropriate at this point to give examples of thesaurus formats currently available in the construction industry. The examples shown here are taken from the *Thesaurus of ERIC Descriptors, Canadian Thesaurus of Construction Science and Technology, and Construction Industry Thesaurus* [13], [14], [11]. Both the ERIC Thesaurus and the Canadian Thesaurus draw upon *The Rules for Updating and Preparing Thesauri* prepared by the Engineers Joint Council [12]. Note the similarity between the two, in contrast to the third, from England.

Examples of Thesauri Formats

U.S.: *Thesaurus of ERIC Descriptors*

ARCHITECTURAL DRAFTING

   BT   DRAFTING

   RT   ARCHITECTURAL
           EDUCATION
           ARCHITECTURAL
           PROGRAMING
           ARCHITECTURE
           BUILDING PLANS

Canada: *Canadian Thesaurus of Construction Science and Technology.*

DRAWINGS

   UF   ENGINEERING DRAWINGS
           GEOMETRICAL DRAWINGS
   BT   CONTRACT DOCUMENTS
           DOCUMENTATION
           (DOCUMENTS)
   NT   DESIGN DRAWINGS
           DETAILS (DRAWINGS)
           DIAGRAMMATIC SKETCHES
           ELEVATIONS (DRAWINGS)
           PERSPECTIVES (DRAWINGS)
           PLANS (DRAWINGS)
           PRODUCTION DRAWINGS
           SECTIONS (DRAWINGS)
           SKETCHES

England: *Construction Industry Thesaurus.*

F13520	Project documentation
F13530	Programmes = Timetables
F13535	Plans of work
F13540	Job manuals
F13550	Site documentation
F13650	Briefs
F13660	Variation orders
F13670	**Drawings** = Architectural drawings
F13680	Standard detail drawings
F13690	Annotated drawings
F13700	Design drawings
F13710	Sketch drawings
F13810	Working drawings = Production drawings
F13830	Location drawings
F13850	Block plans
F13870	Site plans = Layout drawings
F13890	General location drawings

F13990	Component drawings
F14010	Component range drawings
F14030	Component detail drawings
F14130	Assembly drawings

□

It is especially important for a thesaurus to control the use of synonymous terms. Abbreviations should be generally avoided as well because they may not be understood. If it is absolutely necessary to use an abbreviation, it should be treated as a synonym and cross-referenced. Acronyms or initial letter abbreviations are entered in all uppercase letters. This is in addition to a written-out version. However, caution must be used to avoid confusion. Acronyms that are not in general usage will complicate any search. They should not be used as major identifiers and should be written out when they are used for any major designation.

These guidelines, in conjunction with a unified construction language, will provide a basis for the professional seeking to develop a detail storage and retrieval system. Research has shown that the success of a system is dependent on the language of the system [8]. More precise control of vocabulary will lead to the construction of an effective thesaurus.

The example descriptors shown in Section 4-5.1 were selected from the *ERIC Thesaurus* to aid design professionals in developing a national or in-house construction language. These construction related terms were selected and processed during the 1967–1970 operation of a Clearinghouse on Educational Facilities Design. Each descriptor was reviewed, analyzed, and evaluated for acceptance as a "standard" or "commonly used" construction term. An example of the family relationships established for selected terms is identified in the Mini-Thesaurus shown in Section 4-5.2. These terms should be reevaluated for potential use in the detail banking system.

4-5.1 ERIC descriptors appropriate for construction language.

A

ACOUSTICAL ENVIRONMENT
ACOUSTIC INSULATION
ACOUSTICS
AIR CONDITIONING
AIR STRUCTURES
ARCHITECTURAL BARRIERS
ARCHITECTURAL CHARACTER
ARCHITECTURAL PROGRAMMING
ASPHALTS
ATHLETIC FIELDS
AUDIO VIDEO LABORATORIES
AUDIO VISUAL CENTERS
AUDITORIUMS

B

BEHAVIOR
BIDS
BUDGETING
BRICKLAYERS
BRICKLAYING
BUILDING CONVERSION
BUILDING DESIGN
BUILDING EQUIPMENT
BUILDING MATERIALS
BUILDING PLANS
BUILDINGS

C

CAMPUS PLANNING
CARPETING
CEILINGS
CHALKBOARDS
CHIMNEYS
CHURCHES
CITY PLANNING
CLASSROOM DESIGN
CLASSROOM ENVIRONMENT
CLASSROOM FURNITURE
CLASSROOMS
CLIMATE CONTROL
COLLEGE BUILDINGS
COLLEGE HOUSING
COLLEGE PLANNING
COLOR
COLOR PLANNING
COMPONENT BUILDING SYSTEMS
COMPUTER BASED LAB
CONSTRUCTION (PROCESS)
CONSTRUCTION COSTS
CONSTRUCTION NEEDS
CONTRACTS
CONTROLLED ENVIRONMENT
CORRIDORS
COST EFFECTIVENESS
CRITICAL PATH METHOD

D

DAYLIGHT
DESIGN
DESIGN NEEDS
DINING FACILITIES
DISPLAY PANEL
DOORS
DORMITORIES
DRIVEWAYS

E

EDUCATIONAL ENVIRONMENT
EDUCATIONAL FACILITIES
EDUCATIONAL SPECIFICATIONS
ELECTRICAL APPLIANCES
ELECTRICAL SYSTEMS
ELECTRONIC CLASSROOMS
ENGINEERING
ENVIRONMENT
ENVIRONMENTAL CRITERIA
EQUIPMENT
EQUIPMENT MANUFACTURERS
EQUIPMENT STANDARD
EXHAUSTING

F

FACILITIES
FACILITY CASE STUDIES
FACILITY EXPANSION
FACILITY GUIDELINES
FACILITY IMPROVEMENT
FACILITY INVENTORY
FACILITY REQUIREMENTS
FIELD HOUSES
FINISHING
FIRE PROTECTION
FLEXIBLE CLASSROOMS
FLEXIBLE FACILITIES
FLEXIBLE LIGHTING DESIGN
FLOORING
FOOD HANDLING FACILITY
FOOD STORES
FURNITURE
FURNITURE ARRANGEMENT
FURNITURE DESIGN

G

GLARE
GLASS WALLS
GOVERNMENT LIBRARIES
GREENHOUSES
GYMNASIUMS

H

HANDICAPPED
HEALTH FACILITIES

HEATING
HIGH SCHOOL DESIGN
HOSPITALS
HOTELS
HOUSING
HUMIDITY

I

ILLUMINATION LEVELS
INFORMATION PROCESSING
INFORMATION RETRIEVAL
INFORMATION STORAGE
INFORMATION SYSTEMS
INSTITUTIONAL LIBRARIES
INTERIOR DESIGN
INTERIOR SPACE

J

JUNIOR COLLEGES
JUNIOR HIGH SCHOOLS

K

(NONE)

L

LABORATORIES
LABORATORY EQUIPMENT
LANDSCAPING
LAW LIBRARIES
LAW SCHOOLS
LIBRARIES
LIBRARY EQUIPMENT
LIBRARY PLANNING
LIGHT
LIGHTING
LIGHTING DESIGN

M

MASONRY
MEDICAL LIBRARIES
MEDICAL SCHOOLS
MODULAR BUILDING DESIGN
MOVABLE PARTITIONS
MULTIPURPOSE CLASSROOMS
MUSEUMS
MUSIC FACILITIES

N

NEIGHBORHOOD SCHOOLS
NURSERY SCHOOLS

O

OFFICE (FACILITIES)
OPEN PLAN SCHOOLS
OUTDOOR THEATERS

P

PARK DESIGN
PARKING AREAS
PARKING CONTROLS
PARKING FACILITIES
PARKS
PERFORMANCE CRITERIA
PERFORMANCE SPECIFICATIONS
PHYSICAL EDUCATION FACILITIES
PHYSICALLY HANDICAPPED
PLANNING
PLAYGROUNDS
PLUMBING
PREFABRICATION
PRESTRESSED CONCRETE
PUBLIC HOUSING
PUBLIC LIBRARIES

Q

(NONE)

R

RECREATIONAL FACILITIES
REGIONAL PLANNING
RESEARCH LIBRARIES
RESIDENTIAL SCHOOLS
ROOFING

S

SANITARY FACILITIES
SCHOOL BUILDINGS
SCHOOL CONSTRUCTION
SCHOOL DESIGN
SCHOOL LOCATION
SCHOOL PLANNING
SCIENCE LABORATORIES
SEALERS
SECONDARY SCHOOLS
SITE ANALYSIS
SITE DEVELOPMENT
SITE SELECTION
SPACE CLASSIFICATIONS
SPACE DIVIDERS
SPACE UTILIZATION
SPATIAL RELATIONSHIP
SPECIFICATIONS
STAGES
STRUCTURAL BUILDING SYSTEM
SWIMMING POOLS

T

TEMPERATURE
THEATERS
THERMAL ENVIRONMENT
TOILET FACILITIES
TRAFFIC CIRCULATION
TRAFFIC CONTROL
TRAFFIC PATTERNS
TRAFFIC SIGNS
TRANSPORTATION
TREES

U

UNIVERSITIES
UNIVERSITY LIBRARIES
URBAN ENVIRONMENT
URBAN RENEWAL
URBAN SCHOOLS
UTILITIES

V

VEHICULAR TRAFFIC
VENTILATION
VERTICAL WORK SURFACES
VISUALLY HANDICAPPED

W

WATER POLLUTION CONTROL
WINDOWLESS ROOMS
WINDOWS
WOODWORKING
WORK ENVIRONMENT

X

(NONE)

Y

(NONE)

Z

ZONING

4-5.2 Selected ERIC descriptors demonstrating a mini-thesaurus.

A
ACOUSTIC INSULATION

    UF   ACOUSTIC BARRIERS
               ACOUSTIC INSULATORS

ANECHOIC MATERIALS
SOUND ABSORBING MATERIALS
SOUND BARRIERS
SOUND INSULATION
SOUNDPROOFING
SOUND REFLECTING MATERIALS
RT   ACOUSTICAL ENVIRONMENT
ACOUSTICS
BUILDING DESIGN
BUILDING MATERIALS
CONSTRUCTION (PROCESS)
CONTROLLED ENVIRONMENT

AUDITORIUMS

BT   EDUCATIONAL FACILITIES
RT   ARTS CENTERS
MUSIC FACILITIES
STAGES
THEATERS
WINDOWLESS ROOMS

B

BUILDING EQUIPMENT

NT   AIR CONDITIONING EQUIPMENT
FURNITURE
BT   EQUIPMENT
RT   BUILDINGS
CHIMNEYS
EDUCATIONAL EQUIPMENT
EQUIPMENT MANUFACTURERS
MECHANICAL EQUIPMENT
SANITARY FACILITIES
SHEET METAL WORK
STORAGE

BUILDING MATERIALS

UF   CONSTRUCTION MATERIALS
NT   ADHESIVES
ASPHALTS
MASONRY
PRESTRESSED CONCRETE
SEALERS
RT   ACOUSTIC INSULATION
ARCHITECTURAL ELEMENTS
ARCHITECTURAL RESEARCH
BUILDING DESIGN
BUILDINGS
CARPETING
CEMENT INDUSTRY
CONSTRUCTION (PROCESS)
CONSTRUCTION COSTS
CONSTRUCTION NEEDS
DOORS
FACILITIES
FLOORING
PREFABRICATION
ROOFING

C

CLASSROOM FURNITURE

UF   FURNITURE (CLASSROOM)
BT   FURNITURE
RT   CLASSROOMS
FURNITURE ARRANGEMENT
FURNITURE DESIGN
STORAGE

COMPONENT BUILDING SYSTEMS

SN   INTERACTING OR INTERDEPENDENT
STRUCTURAL OR MECHANICAL BUILDING
ELEMENTS DESIGNED AND CONSTRUCTED
IN TERMS OF FLEXIBILITY AND ECONOMICS
UF   COMPONENT SYSTEMS
SYSTEM COMPONENTS
RT   ARCHITECTURAL RESEARCH
ARCHITECTURE
CONSTRUCTION COSTS
CONSTRUCTION NEEDS
FACILITY GUIDELINES
FLEXIBLE CLASSROOMS
FLEXIBLE FACILITIES
MASTER PLANS
MODULAR BUILDING DESIGN
PLANNING
PREFABRICATION
PRESTRESSED CONCRETE
SCHOOL CONSTRUCTION
SCHOOL DESIGN
SCHOOL PLANNING
STANDARDS
STRUCTURAL BUILDING SYSTEMS

D

DINING FACILITIES

UF   CAFETERIAS
DINING ROOMS
SNACK BARS
BT   FACILITIES
RT   DISHWASHING
FOOD HANDLING FACILITIES
FOOD SERVICE

DOORS

BT   ARCHITECTURAL ELEMENTS
RT   BUILDING DESIGN
BUILDING MATERIALS

E

ELECTRICAL SYSTEMS

UF   ELECTRIC SYSTEMS
RT   BUILDING DESIGN
ELECTRICIANS
ELECTRICITY
ELECTRONIC CONTROL
ELECTRONICS

LIGHTING
UTILITIES

EQUIPMENT STANDARDS

BT   STANDARDS

RT   EQUIPMENT
EQUIPMENT EVALUATION
EQUIPMENT MANUFACTURERS
FIELD CHECK
MECHANICAL EQUIPMENT
PERFORMANCE SPECIFICATIONS
PURCHASING
SPECIFICATIONS

F

FLEXIBLE LIGHTING DESIGN

SN   LIGHTING UNIT ARRANGEMENT AS WELL AS
LIGHTING FIXTURE DESIGN THAT
ALLOWS FOR FLEXIBLE LIGHTING
REQUIREMENTS

BT   DESIGN
LIGHTING DESIGN

RT   BUILDING DESIGN
CLASSROOM ARRANGEMENT
FLEXIBLE FACILITIES
ILLUMINATION LEVELS
LIGHTING
LIGHTS
MULTIPURPOSE CLASSROOMS

FURNITURE DESIGN

BT   DESIGN

RT   CLASSROOM FURNITURE
DESIGN NEEDS
DESIGN PREFERENCES
FURNITURE
FURNITURE ARRANGEMENT
FURNITURE INDUSTRY
LUMBER INDUSTRY
PHYSICAL DESIGN NEEDS
PSYCHOLOGICAL DESIGN NEEDS
STORAGE

G

GLASS WALLS

SN   WALLS CONSISTING LARGELY OF WINDOWS

UF   WINDOW WALLS

RT   BUILDING DESIGN
CLASSROOM DESIGN
SCHOOL DESIGN
WINDOWS

GYMNASIUMS

BT   PHYSICAL EDUCATION FACILITIES

RT   ATHLETIC ACTIVITIES
ATHLETIC EQUIPMENT
ATHLETIC FIELDS

ATHLETIC PROGRAMS
ATHLETICS

H

HIGH SCHOOL DESIGN

BT   SCHOOL DESIGN

RT   HIGH SCHOOLS

HOSPITALS

UF   SANATORIUMS

NT   PSYCHIATRIC HOSPITALS

BT   HEALTH FACILITIES

RT   CLINICS
HEALTH SERVICES
HOSPITAL SCHOOLS
INSTITUTION LIBRARIES
MEDICAL SERVICES
NURSING HOMES

I

ILLUMINATION LEVELS

SN   LIGHTING REQUIREMENTS PRIMARILY
MEASURED IN FOOT CANDLES

UF   FOOT CANDLES

RT   COLOR PLANNING
CONTRAST
DAYLIGHT
FLEXIBLE LIGHTING DESIGN
GLARE
LIGHT
LIGHTING
LIGHTING DESIGN
LIGHTS
LUMINESCENCE
OPTICS
PHYSICAL ENVIRONMENT
TASK PERFORMANCE
VISUAL DISCRIMINATION
VISUAL PERCEPTION
WINDOWS

INTERIOR DESIGN

UF   INTERIOR DECORATING
INTERIOR DECORATION

BT   DESIGN

RT   ACOUSTICAL ENVIRONMENT
ARCHITECTURE
BUILDING DESIGN
CLASSROOM DESIGN
COLOR PLANNING
DESIGN NEEDS
DESIGN PREFERENCES
FURNITURE ARRANGEMENT
INTERIOR SPACE
LIGHTING
LIGHTING DESIGN
OFFICES (FACILITIES)

PHYSICAL DESIGN NEEDS
PHYSICAL ENVIRONMENT
PSYCHOLOGICAL DESIGN NEEDS
SPACE CLASSIFICATION
SPACE UTILIZATION
SPATIAL RELATIONSHIP
THERMAL ENVIRONMENT

J

JUNIOR COLLEGES

BT   COLLEGES

RT   ASSOCIATE DEGREES
COMMUNITY COLLEGES
HIGHER EDUCATION
JUNIOR COLLEGE LIBRARIES
JUNIOR COLLEGE STUDENTS
POST SECONDARY EDUCATION
STATE COLLEGES
TECHNICAL INSTITUTES
UNDERGRADUATE STUDY

JUNIOR HIGH SCHOOLS

BT   HIGH SCHOOLS

RT   JUNIOR HIGH SCHOOL STUDENTS
MIDDLE SCHOOLS
SENIOR HIGH SCHOOLS

K
(NONE)

L

LANDSCAPING

BT   ORNAMENTAL HORTICULTURE

RT   FLORICULTURE
GROUNDS KEEPERS
PLANT IDENTIFICATION
PLANT SCIENCE
SITE DEVELOPMENT
TRAILS
TURF MANAGEMENT

LIGHTING DESIGN

NT   FLEXIBLE LIGHTING DESIGN

BT   DESIGN

RT   ARCHITECTURE
BUILDING DESIGN
CONTRAST
DESIGN NEEDS
GLARE
ILLUMINATION LEVELS
INTERIOR DESIGN
LIGHTING
LIGHTS
WINDOWLESS ROOMS

M

MASONRY

BT   BUILDING MATERIALS

RT   ARCHITECTURAL ELEMENTS

BRICKLAYING
CONSTRUCTION (PROCESS)
PREFABRICATION
PRESTRESSED CONCRETE
STRUCTURAL BUILDING SYSTEMS

MOVABLE PARTITIONS

SN   INTERIOR WALLS THAT CAN BE READILY
MOVED

UF   FOLDING PARTITIONS

BT   SPACE DIVIDERS

RT   FLEXIBLE CLASSROOMS
FLEXIBLE FACILITIES
PREFABRICATION
SPACE UTILIZATION

N

NEIGHBORHOOD SCHOOLS

BT   SCHOOLS

RT   COMPENSATORY EDUCATION
NEIGHBORHOOD
NEIGHBORHOOD SCHOOL POLICY

NURSERY SCHOOLS

BT   SCHOOLS

RT   CHILD CARE WORKERS
PRESCHOOL EDUCATION

O

OFFICES (FACILITIES)

UF   FACULTY OFFICES
STAFF OFFICES

BT   FACILITIES

RT   INTERIOR DESIGN
INTERIOR SPACE
SCHOOL SPACE
SPACE CLASSIFICATION

OUTDOOR THEATERS

UF   AMPHITHEATERS
OPEN AIR THEATERS

BT   THEATERS

RT   DRAMATICS
OUTDOOR DRAMA
STAGES
THEATER ARTS

P

PARKING FACILITIES

SN   ABOVE AND/OR BELOW GROUND
STRUCTURES FOR STORAGE OF VEHICLES

UF   PARKING GARAGES
PARKING RAMPS
UNDERGROUND GARAGES

NT   BUS GARAGES

BT   FACILITIES

RT CAMPUS PLANNING
      DRIVEWAYS
      MOTOR VEHICLES
      PARKING CONTROLS
      VEHICULAR TRAFFIC

PRESTRESSED CONCRETE

UF POST TENSIONED CONCRETE
      PRETENSIONED CONCRETE

BT BUILDING MATERIALS

RT ARCHITECTURAL ELEMENTS
      CEMENT INDUSTRY
      COMPONENT BUILDING SYSTEMS
      MASONRY
      PREFABRICATION
      STRUCTURAL BUILDING SYSTEMS

Q
(NONE)

R
RECREATIONAL FACILITIES

NT COMMUNITY ROOMS
      PARKS
      PLAYGROUNDS
      SWIMMING POOLS
      ZOOS

BT FACILITIES

RT COMMUNITY RESOURCES
      FIELD HOUSES
      PHYSICAL EDUCATION FACILITIES
      RECREATION
      RECREATION FINANCES
      STUDENT UNIONS
      TRAILS
      WINDOWLESS ROOMS

ROOFING

UF ROOF COVERING
      ROOF INSTALLATION
      ROOFS

RT ASPHALTS
      BUILDING MATERIALS
      BUILDINGS
      BUILDING TRADES
      CONSTRUCTION (PROCESS)
      ROOFERS

S
SPACE CLASSIFICATION

SN CATEGORIZATION OF AREAS IN A GIVEN
      FACILITY GENERALLY BY FUNCTION OR
      PURPOSE

BT CLASSIFICATION

RT BUILDING PLANS
      COLLEGE PLANNING
      DESIGN NEEDS
      FACILITY INVENTORY

      FACILITY UTILIZATION RESEARCH
      INTERIOR DESIGN
      INTERIOR SPACE
      OFFICES (FACILITIES)
      SPACE UTILIZATION
      SPATIAL RELATIONSHIP

STRUCTURAL BUILDING SYSTEMS

SN COMBINATION OF SUCH STRUCTURAL
      MEMBERS AND METHODS AS
      FOUNDATIONS, POST AND BEAM, VAULTS,
      OR LIFT-SLABS TO FORM THE STRUCTURAL
      FRAME OR SHELL OF A BUILDING

RT ARCHITECTURAL RESEARCH
      ARCHITECTURE
      BUILDING DESIGN
      BUILDINGS
      CIVIL ENGINEERING
      COMPONENT BUILDING SYSTEMS
      CONSTRUCTION (PROCESS)
      CONSTRUCTION COSTS
      MASONRY
      PREFABRICATION
      PRESTRESSED CONCRETE
      SCHOOL ARCHITECTURE
      SCHOOL BUILDINGS
      SCHOOL CONSTRUCTION
      SCHOOL DESIGN

T
TOILET FACILITIES

BT SANITARY FACILITIES

RT HEALTH FACILITIES
      PUBLIC FACILITIES

TRAFFIC CIRCULATION

UF TRAFFIC FLOW

RT COMMUTING STUDENTS
      DRIVEWAYS
      MOTOR VEHICLES
      PARKING AREAS
      PARKING CONTROLS
      PEDESTRIAN TRAFFIC
      ROAD CONSTRUCTION
      SERVICE VEHICLES
      STUDENT TRANSPORTATION
      TRAFFIC CONTROL
      TRAFFIC PATTERNS
      TRAFFIC SIGNS
      VEHICULAR TRAFFIC

U
UNIVERSITIES

NT LAND GRANT UNIVERSITIES
      STATE UNIVERSITIES
      URBAN UNIVERSITIES

BT INSTITUTIONS

RT COLLEGE BUILDINGS
      COLLEGES

ENGLISH DEPARTMENTS
EXPERIMENTAL COLLEGES
EXTENSION EDUCATION
GRADUATE STUDY
HIGHER EDUCATION
PRIVATE COLLEGES
SCIENCE DEPARTMENTS
TEACHING ASSISTANTS
UNDERGRADUATE STUDY
UNIVERSITY LIBRARIES

UTILITIES

UF   ELECTRIC UTILITIES
GAS UTILITIES
PUBLIC UTILITIES
WATER UTILITIES

BT   SERVICES

RT   COMMUNICATIONS
ELECTRICAL SYSTEMS
FUELS
HEATING
KINETICS
LIGHTING
SANITARY FACILITIES
SANITATION
TELEPHONE COMMUNICATION SYSTEMS

V
VEHICULAR TRAFFIC

UF   VEHICULAR CIRCULATION

RT   CAMPUS PLANNING
DRIVEWAYS
MASTER PLANS
MOTOR VEHICLES
PARKING AREAS
PARKING CONTROLS
PARKING FACILITIES
PEDESTRIAN TRAFFIC
SERVICE VEHICLES
TRAFFIC CIRCULATION
TRAFFIC CONTROL
TRAFFIC PATTERNS
TRAFFIC REGULATIONS
TRAFFIC SIGNS
TRANSPORTATION

VENTILATION

RT   AIR CONDITIONING
AIR CONDITIONING EQUIPMENT
AIR FLOW
CHIMNEYS
CLIMATE CONTROL
CONTROLLED ENVIRONMENT
DESIGN NEEDS
EXHAUSTING
FUEL CONSUMPTION
HEATING
LIGHTING
MECHANICAL EQUIPMENT
PHYSICAL ENVIRONMENT

TEMPERATURE
THERMAL ENVIRONMENT
WINDOWLESS ROOMS
WINDOWS

W
WINDOWLESS ROOMS

SN   ANY AREA IN A BUILDING CLOSED TO
EXTERIOR ENVIRONMENT

BT   FACILITIES

RT   AIR CONDITIONING
AUDITORIUMS
CLIMATE CONTROL
CORRIDORS
FALLOUT SHELTERS
LIGHTING
LIGHTING DESIGN
RECREATIONAL FACILITIES
VENTILATION

WINDOWS

UF   FENESTRATION

RT   BUILDING DESIGN
CLIMATE CONTROL
DAYLIGHT
GLARE
GLASS WALLS
ILLUMINATION LEVELS
LIGHTING
VENTILATION

X
(NONE)

Y
(NONE)

Z
ZONING

NT   COMMUNITY ZONING
REZONING
SCHOOL ZONING
SPECIAL ZONING

RT   MULTICAMPUS DISTRICTS
REAL ESTATE
SCHOOL DISTRICTS

□

4-6   DIRECTIVES FOR CREATING A LANGUAGE AND A THESAURUS

A large percent of the terms required for detail banking already exist in the following documents:

▷ *Uniform Construction Index* [10].
▷ *Thesaurus of ERIC Descriptors* [13], [15].

▷ *Thesaurus of Engineering and Scientific Terminology* [16].

▷ *HUD Research Thesaurus* [19].

▷ *Canadian Thesaurus of Construction Science and Technology* [11].

▷ *English Construction Industry Thesaurus* [14].

Many of the terms identified in the listed documents also fulfill the requirements for structuring a construction language. Initial efforts should be directed toward using generic terms already evaluated and accepted in the construction field. Coordinating these terms will save time, money, and research effort in creating a major portion of the construction language thesaurus.

An efficient information storage and retrieval system must have terminology control at both the indexing stage and the searching stage. Therefore every effort should be made to create a construction language that can be used nationwide. However, to not delay development of effective information-handling programs, it may be necessary for organizations to select and coordinate standard terms for in-house detail banking systems. The following directives should be considered in developing the language and thesaurus.

4-6.1 Language preparation directives.

1. Descriptor terms should be selected from a common language used in construction.

2. Select established standard construction terms generated by national organizations, associations, and building research centers before creating a new language. Consult AIA, CSI, NSPE, and other design-related organizations for existing standardized terminology.

3. Words should be selected to represent subject areas required to describe each construction detail.

4. Terminology should be universally accepted by all disciplines in the construction field.

5. Broad and narrow terms should be cross-referenced with major subject areas to create

family relationships that aid information indexing and retrieval.

6. One language should be created to store and retrieve all information processed for the detail bank.

7. A construction thesaurus should be structured to control the terms used in detail banking.

4-6.2 Thesaurus preparation directives.

1. Appoint or select a construction language control committee to share the responsibility of making recommendations, evaluations, and final decisions.

2. Select a manager or leader to take charge of program organization and administration.

3. Direct a researcher to review and analyze construction documents (working drawings and specifications) for commonly reused terms that have potential as a construction vocabulary.

4. Reference existing rules and guidelines for selecting construction terminology. (See references cited in step 5 at the beginning of Secton 6-2).

5. Develop a procedure for selecting and evaluating descriptor terms for the construction language.

   a. Appoint individual to prepare terms for review and evaluation.

   b. Evaluate terms by utilizing a committee review process.

   c. Conduct evaluation reviews as required to select new or questionable terms.

   d. Approve terms to be included in thesaurus.

6. Identify descriptor terms that are frequently reused on working drawing details.

7. Select acceptable descriptor terms already used by associations and organizations in the construction industry.

8. Search existing thesauri for additional terms to be used in construction thesaurus.

9. Identify new descriptor terms when required to describe detail content.

10. Evaluate potential construction terms by using reference documents, dictionaries, and existing thesauri to provide support for their selection. (See references cited in Section 4-4.2.)

11. Structure the thesaurus with approved construction vocabulary.

12. Develop the framework for each descriptor to be part of a generic or family relationship of (1) broad, (2) narrow, and (3) related terms.

13. List terms alphabetically as they are prepared and added to the construction thesaurus.

14. Prepare descriptor scope notes for potentially confusing terminology. (See guidelines taken from *Rules for Thesaurus Preparation in Section 6-1.*)

15. Do not confuse language and thesaurus development with the simplicity of using construction vocabulary for storing and retrieving information. *These two tasks are completely different activities that must be programmed separately. Once the construction language is developed, which could be on a national level, the procedures for indexing, storing, and retrieving information are very simple and direct.*

FIVE
EVALUATING INFORMATION-HANDLING SYSTEMS

5-1 BANKING SYSTEM REQUIREMENTS

Every detail banking system should be designed to meet specific requirements as determined by office needs and objectives. These needs should be based on facility design requirements, internal office structure, and future projections of the firm. Anticipated annual growth of a detail banking system should be predetermined before structuring the system for handling working drawing details. For example, if the firm is going to remain small and work with limited facility types, a manual or semiautomated system may be structured to handle such needs. Whereas in a firm with large anticipated growth, it is recommended that the initial system design and planning incorporate flexibility to work with automation. Automation will help larger firms control a greater volume of working drawing details.

The storage and retrieval process used in detail banking is the most difficult program to select and develop because a variety of options exist. Therefore a specific plan must be devised to store construction details in an effective and systematic manner. It takes time and effort to construct an appropriate indexing system, but an evaluation of potential storage systems that fulfill the firm's specific needs is essential. The following features should be considered critical. An effective detail banking system will:

▷ Group and relate details stored in different categories.

▷ Cross-reference details by building material or detail types.

▷ Obtain particular details with the greatest ease and in the shortest possible time.

▷ Use a storage and retrieval language consistent with building design and construction activities.

▷ Have a simple retrieval system that draftspersons and other design team members can understand with minimum training.

5-1.1 System selection.

Determining the best indexing system is perhaps the most difficult part of planning an effective detail bank. A wide variety of retrieval systems are presently in use—numerical, alphanumeric, key word, subject area descriptors, specification and section division numbering, among other combinations. The following list identifies many of the classification systems presently in use.

Classification, Numbering, and Cataloging Systems

1. Numerical systems.
2. Alphanumerical systems.
3. Descriptor terms/thesauri—the recommended system.
4. Key word terms.
5. Color coding.
6. Author index.
7. Product index.
8. Manufacturer's index.
9. Uniform Construction Index (UCI)—organizes product related information into 16 broad divisions.
10. *Construction Industry Thesaurus* (England).
11. *Thesaurus of ERIC Descriptors* (U.S. HEW).

12. *Canadian Thesaurus of Construction Science and Technology* (Canada).

13. Library of Congress.

14. Dewey Decimal System—assigns specific numbers to each document for filing and retrieval.

15. Universal Decimal Classification System—classification of information into 10 areas
    Example:

Section Code:	6	Applied Science
Sub Section:	69	Building
	691	Building Materials
	691.1	Organic Building Materials

The Descriptor Term System can be developed from the standard construction language used in systems numbered 4, 9, 10, 11, and 12. Only systems structured with a construction language can be accepted for a detail banking thesaurus.

Information-handling requirements should be the prime determinants in the system selection process. The important point to remember is that system selection be consistent with the overall plan for information handling within an organization. *A major objective should be to eliminate developing complex retrieval systems that confuse the user. When selecting a system, consider one:*

▷ That is open-ended and flexible so that new details and construction-related information can be added.

▷ That allows the indexer flexibility in selecting the appropriate retrieval category. (Systems that provide too many retrieval category options will create confusion for the indexer in determining the most appropriate category for storage.)

▷ That can be utilized in storing other general information handled by the design firm. A data base that can work for details and other categories of information is most desirable, since it will simplify the firm's storage and retrieval system.

One person should be put in charge of managing the detail banking system so that users can obtain information quickly and easily without having to perform individual searches. The person maintaining the filing system must organize an indexing plan consistent with the firm's facility design and construction program. By centralizing system control, a firm can increase efficiency and reduce the time required to retrieve desired information. To begin the process, the following steps should be taken:

1. Identify critical goals and objectives for your detail banking plan.

2. Develop a detail banking plan of action.

3. Select a person with information-handling training to take charge of managing the detail banking system.

4. Research existing information-handling and indexing systems as a basis for pinpointing specific detail banking goals.

5. Evaluate existing systems in terms of your specific needs and special requirements.

6. Select a detail indexing system that is flexible to handle a variety of expanding building systems and is consistent with your overall information-handling plan.

7. Choose and develop a system that is easy to understand, manage, and use.

5-2   TYPES OF INDEXING SYSTEMS

A variety of storage and retrieval systems are presently being experimented with and used by some firms. However, there appears to be no one system that has great potential for effectively handling the detail banking system. This chapter is structured to explore these systems and highlight the advantages and disadvantages so that a master system can emerge for detail banking. The learning experiences of several firms are cited to identify the critical problem areas associated with existing information systems. Some of these problems can be summarized as follows:

▷ It is difficult to determine which division, section, or category to file under.

▷ When a specific location is selected, it is not one that somebody else will be able to use to find the detail.

▷ The numbering and coding system is too difficult to understand.

▷ The filing system is limited to specific categories.

▷ The system is not managed or updated.

In selecting an appropriate information system, it is important to overcome past problems. Most existing systems have a limited range of application and therefore cause users frustration as they expand to include broader areas of information. Several variations of the following systems will be identified and evaluated for use in detail banking:

1. The Uniform Construction Index System (CSI) [10].

2. The English Construction Industry Thesaurus System [14].

3. Alphanumeric systems

5-2.1 Uniform Construction Index.

The *Uniform Construction Index* was developed through a joint effort of the U.S. and Canadian construction industries to standardize a system for classification and retrieval of technical data. Terminology based on specifications was selected to identify place, trade, function, and materials inherent in construction. The Construction Specifications Institute (CSI) Format for Building Specifications is used to group terms into 16 Divisions. Each Division is a broad generic grouping of related units of work. A unit of work is referred to as a section that describes a material or product and its installation. Major sections of a Division are referred to as "Broadscope Sections" while smaller and more specific sections are called "Narrowscope." The Division-section format provides appropriate headings for filing and retrieving information. To aid the indexing process, a five-digit numbering system is used for designating all sections and divisions. An example of

the CSI Division-Section Format is shown in Figures 5-1 and 5-2 [10].

A Key Word Index has also been developed to organize section titles and classifications alphabetically for ease of reference. Each key word is number referenced to the appropriate Division and Broadscope heading. This arrangement of the terms and numbering system provides the indexer and researcher with a means of storing and retrieving information. An example of the Key Word Index is shown in Figure 5-3.

To adapt the *Uniform Construction Index* system for details requires expanding the numbers to identify each detail as it is added to the bank. The following procedures can be used to develop a banking system linked to construction specifications.

Detail Filing Using UCI Format

1. Each detail is filed under the Divisions of the CSI Format.

2. Details are given numbers associated with the key word subject area.
   Example: A. Metal Decking Details are given a 05300 number.
   B. Handrails and Railing Details are given a 05520 number.
   C. If more than one detail is collected in a section, a three- or four-digit number can be added: 05300-0001.

3. Details can be filed numerically in notebooks for each Division and Section.

4. Key words, specifications, and details can also be linked for storage and retrieval with word processors and computers.

The examples shown in Figures 5-4, 5-5, 5-6, 5-7, and 5-8 demonstrate development and use of variations in the *Uniform Construction Index* System.

▷ Example 1: Figure 5-4.
▷ Example 2: Figures 5-5, 5-6, and 5-7.
▷ Example 3: Figure 5-8.

DIVISION 0 – BIDDING AND CONTRACT REQUIREMENTS

00010	PRE-BID INFORMATION
00100	INSTRUCTIONS TO BIDDERS
00200	INFORMATION AVAILABLE TO BIDDERS
00300	BID/TENDER FORMS
00400	SUPPLEMENTS TO BID/TENDER FORMS
00500	AGREEMENT FORMS
00600	BONDS AND CERTIFICATES
00700	GENERAL CONDITIONS OF THE CONTRACT
00800	SUPPLEMENTARY CONDITIONS
00950	DRAWINGS INDEX
00900	ADDENDA AND MODIFICATIONS

SPECIFICATIONS—DIVISIONS 1-16

DIVISION 1 – GENERAL REQUIRMENTS

01010	SUMMARY OF WORK
01020	ALLOWANCES
01030	SPECIAL PROJECT PROCEDURES
01040	COORDINATION
01050	FIELD ENGINEERING
01060	REGULATORY REQUIRMENTS
01070	ABBREVIATIONS AND SYMBOLS
01080	IDENTIFICATION SYSTEMS
01100	ALTERNATES/ALTERNATIVES
01150	MEASUREMENT AND PAYMENT
01200	PROJECT MEETINGS
01300	SUBMITTALS
01400	QUALITY CONTROL
01500	CONSTRUCTION FACILITIES AND TEMPORARY CONTROLS
01600	MATERIAL AND EQUIPMENT
01650	STARTING OF SYSTEMS
01660	TESTING, ADJUSTING, AND BALANCING OF SYSTEMS
01700	CONTRACT CLOSEOUT
01800	MAINTENANCE MATERIALS

DIVISION 2 – SITEWORK

02010	SUBSURFACE INVESTIGATION
02050	DEMOLITION
02100	SITE PREPARATION
02150	UNDERPINNING
02200	EARTHWORK
02300	TUNNELLING
02350	PILES, CAISSONS AND COFFERDAMS
02400	DRAINAGE
02440	SITE IMPROVEMENTS
02480	LANDSCAPING
02500	PAVING AND SURFACING
02580	BRIDGES
02590	PONDS AND RESERVOIRS
02600	PIPED UTILITY MATERIALS AND METHODS
02700	PIPED UTILITIES
02800	POWER AND COMMUNICATION UTILITIES
02850	RAILROAD WORK
02880	MARINE WORK

DIVISION 3 – CONCRETE

03010	CONCRETE MATERIALS
03050	CONCRETING PROCEDURES
03100	CONCRETE FORMWORK
03150	FORMS
03180	FORM TIES AND ACCESSORIES
03200	CONCRETE REINFORCEMENT
03250	CONCRETE ACCESSORIES
03300	CAST-IN-PLACE CONCRETE
03350	SPECIAL CONCRETE FINISHES
03360	SPECIALLY PLACED CONCRETE
03370	CONCRETE CURING
03400	PRECAST CONCRETE
03500	CEMENTITIOUS DECKS
03600	GROUT
03700	CONCRETE RESTORATION AND CLEANING

DIVISION 4 – MASONRY

04050	MASONRY PROCEDURES
04100	MORTAR
04150	MASONRY ACCESSORIES
04200	UNIT MASONRY
04400	STONE
04500	MASONRY RESTORATION AND CLEANING
04550	REFRACTORIES
04600	CORROSION RESISTANT MASONRY

DIVISION 5 – METALS

05010	METAL MATERIALS AND METHODS
05050	METAL FASTENING
05100	STRUCTURAL METAL FRAMING
05200	METAL JOISTS
05300	METAL DECKING
05400	COLD-FORMED METAL FRAMING
05500	METAL FABRICATIONS
05700	ORNAMENTAL METAL
05800	EXPANSION CONTROL
05900	METAL FINISHES

DIVISION 6 – WOOD AND PLASTICS

06050	FASTENERS AND SUPPORTS
06100	ROUGH CARPENTRY
06130	HEAVY TIMBER CONSTRUCTION
06150	WOOD-METAL SYSTEMS
06170	PREFABRICATED STRUCTURAL WOOD
06200	FINISH CARPENTRY
06300	WOOD TREATMENT
06400	ARCHITECTURAL WOODWORK
06500	PREFABRICATED STRUCTURAL PLASTICS
06600	PLASTIC FABRICATIONS

DIVISION 7 – THERMAL AND MOISTURE PROTECTION

07100	WATERPROOFING
07150	DAMPPROOFING
07200	INSULATION
07250	FIREPROOFING
07300	SHINGLES AND ROOFING TILES
07400	PREFORMED ROOFING AND SIDING
07500	MEMBRANE ROOFING
07570	TRAFFIC TOPPING
07600	FLASHING AND SHEET METAL
07800	ROOF ACCESSORIES
07900	SEALANTS

DIVISION 8 – DOORS AND WINDOWS

08100	METAL DOORS AND FRAMES
08200	WOOD AND PLASTIC DOORS
08250	DOOR OPENING ASSEMBLIES
08300	SPECIAL DOORS
08400	ENTRANCES AND STOREFRONTS
08500	METAL WINDOWS
08600	WOOD AND PLASTIC WINDOWS
08650	SPECIAL WINDOWS
08700	HARDWARE
08800	GLAZING
08900	GLAZED CURTAIN WALLS

DIVISION 9 – FINISHES

09100	METAL SUPPORT SYSTEMS
09200	LATH AND PLASTER
09230	AGGREGATE COATINGS
09250	GYPSUM WALLBOARD
09300	TILE
09400	TERRAZZO
09500	ACOUSTICAL TREATMENT
09550	WOOD FLOORING
09600	STONE AND BRICK FLOORING
09650	RESILIENT FLOORING
09680	CARPETING
09700	SPECIAL FLOORING
09760	FLOOR TREATMENT
09800	SPECIAL COATINGS
09900	PAINTING
09950	WALL COVERING

(a)

Figure 5-1 The CSI MASTERFORMAT. (The Construction Specifications Institute. CSI Document MP-2-1 MASTERFORMAT. Master List of Section Titles and Numbers. 601 Madison Street, Alexandria, Virginia 22314.)

DIVISION 10 – SPECIALTIES

10100	CHALKBOARDS AND TACKBOARDS
10150	COMPARTMENTS AND CUBICLES
10200	LOUVERS AND VENTS
10240	GRILLES AND SCREENS
10250	SERVICE WALL SYSTEMS
10260	WALL AND CORNER GUARDS
10270	ACCESS FLOORING
10280	SPECIALTY MODULES
10290	PEST CONTROL
10300	FIREPLACES AND STOVES
10340	PREFABRICATED STEEPLES, SPIRES, AND CUPOLAS
10350	FLAGPOLES
10400	IDENTIFYING DEVICES
10450	PEDESTRIAN CONTROL DEVICES
10500	LOCKERS
10520	FIRE EXTINGUISHERS, CABINETS, AND ACCESSORIES
10530	PROTECTIVE COVERS
10550	POSTAL SPECIALTIES
10600	PARTITIONS
10650	SCALES
10670	STORAGE SHELVING
10700	EXTERIOR SUN CONTROL DEVICES
10750	TELEPHONE ENCLOSURES
10800	TOILET AND BATH ACCESSORIES
10900	WARDROBE SPECIALTIES

DIVISION 11 – EQUIPMENT

11010	MAINTENANCE EQUIPMENT
11020	SECURITY AND VAULT EQUIPMENT
11030	CHECKROOM EQUIPMENT
11040	ECCLESIASTICAL EQUIPMENT
11050	LIBRARY EQUIPMENT
11060	THEATER AND STAGE EQUIPMENT
11070	MUSICAL EQUIPMENT
11080	REGISTRATION EQUIPMENT
11100	MERCANTILE EQUIPMENT
11110	COMMERCIAL LAUNDRY AND DRY CLEANING EQUIPMENT
11120	VENDING EQUIPMENT
11130	AUDIO-VISUAL EQUIPMENT
11140	SERVICE STATION EQUIPMENT
11150	PARKING EQUIPMENT
11160	LOADING DOCK EQUIPMENT
11170	WASTE HANDLING EQUIPMENT
11190	DETENTION EQUIPMENT
11200	WATER SUPPLY AND TREATMENT EQUIPMENT
11300	FLUID WASTE DISPOSAL AND TREATMENT EQUIPMENT
11400	FOOD SERVICE EQUIPMENT
11450	RESIDENTIAL EQUIPMENT
11460	UNIT KITCHENS
11470	DARKROOM EQUIPMENT
11480	ATHLETIC, RECREATIONAL, AND THERAPEUTIC EQUIPMENT
11500	INDUSTRIAL AND PROCESS EQUIPMENT
11600	LABORATORY EQUIPMENT
11650	PLANETARIUM AND OBSERVATORY EQUIPMENT
11700	MEDICAL EQUIPMENT
11780	MORTUARY EQUIPMENT
11800	TELECOMMUNICATION EQUIPMENT
11850	NAVIGATION EQUIPMENT

DIVISION 12 – FURNISHINGS

12100	ARTWORK
12300	MANUFACTURED CABINETS AND CASEWORK
12500	WINDOW TREATMENT
12550	FABRICS
12600	FURNITURE AND ACCESSORIES
12670	RUGS AND MATS
12700	MULTIPLE SEATING
12800	INTERIOR PLANTS AND PLANTINGS

DIVISION 13 – SPECIAL CONSTRUCTION

13010	AIR SUPPORTED STRUCTURES
13020	INTEGRATED ASSEMBLIES
13030	AUDIOMETRIC ROOMS
13040	CLEAN ROOMS
13050	HYPERBARIC ROOMS
13060	INSULATED ROOMS
13070	INTEGRATED CEILINGS
13080	SOUND, VIBRATION, AND SEISMIC CONTROL
13090	RADIATION PROTECTION
13100	NUCLEAR REACTORS
13110	OBSERVATORIES
13120	PRE-ENGINEERED STRUCTURES
13130	SPECIAL PURPOSE ROOMS AND BUILDINGS
13140	VAULTS
13150	POOLS
13160	ICE RINKS
13170	KENNELS AND ANIMAL SHELTERS
13200	SEISMOGRAPHIC INSTRUMENTATION
13210	STRESS RECORDING INSTRUMENTATION
13220	SOLAR AND WIND INSTRUMENTATION
13410	LIQUID AND GAS STORAGE TANKS
13510	RESTORATION OF UNDERGROUND PIPELINES
13520	FILTER UNDERDRAINS AND MEDIA
13530	DIGESTION TANK COVERS AND APPURTENANCES
13540	OXYGENATION SYSTEMS
13550	THERMAL SLUDGE CONDITIONING SYSTEMS
13560	SITE CONSTRUCTED INCINERATORS
13600	UTILITY CONTROL SYSTEMS
13700	INDUSTRIAL AND PROCESS CONTROL SYSTEMS
13800	OIL AND GAS REFINING INSTALLATIONS AND CONTROL SYSTEMS
13900	TRANSPORTATION INSTRUMENTATION
13940	BUILDING AUTOMATION SYSTEMS
13970	FIRE SUPPRESSION AND SUPERVISORY SYSTEMS
13980	SOLAR ENERGY SYSTEMS
13990	WIND ENERGY SYSTEMS

DIVISION 14 – CONVEYING SYSTEMS

14100	DUMBWAITERS
14200	ELEVATORS
14300	HOISTS AND CRANES
14400	LIFTS
14500	MATERIAL HANDLING SYSTEMS
14600	TURNTABLES
14700	MOVING STAIRS AND WALKS
14800	POWERED SCAFFOLDING
14900	TRANSPORTATION SYSTEMS

DIVISION 15 – MECHANICAL

15050	BASIC MATERIALS AND METHODS
15200	NOISE, VIBRATION, AND SEISMIC CONTROL
15250	INSULATION
15300	SPECIAL PIPING SYSTEMS
15400	PLUMBING SYSTEMS
15450	PLUMBING FIXTURES AND TRIM
15500	FIRE PROTECTION
15600	POWER OR HEAT GENERATION
15650	REFRIGERATION
15700	LIQUID HEAT TRANSFER
15800	AIR DISTRIBUTION
15900	CONTROLS AND INSTRUMENTATION

DIVISION 16 – ELECTRICAL

16050	BASIC MATERIALS AND METHODS
16200	POWER GENERATION
16300	POWER TRANSMISSION
16400	SERVICE AND DISTRIBUTION
16500	LIGHTING
16600	SPECIAL SYSTEMS
16700	COMMUNICATIONS
16850	HEATING AND COOLING
16900	CONTROLS AND INSTRUMENTATION

(b)

Figure 5-1 (Continued)

DIVISION 5—METALS

Number	Title
05010	METAL MATERIALS AND METHODS
05050	METAL FASTENING
05100	STRUCTURAL METAL FRAMING
05200	METAL JOISTS
05300	METAL DECKING
05400	COLD-FORMED METAL FRAMING
05500	METAL FABRICATIONS
05700	ORNAMENTAL METAL
05800	EXPANSION CONTROL
05900	METAL FINISHES

The Construction Specifications Institute
CSI Document MP—2—1 MASTERFORMAT
—Master List of Section Titles and Numbers

DIVISION 5 - METALS

NUMBER	TITLE
05010	METAL MATERIALS AND METHODS
-11	Stainless Steel
-12	Bronze
-13	Aluminum
05050	METAL FASTENING
-60	Welding
-70	Bolting
-80	Riveting
05100	STRUCTURAL METAL FRAMING
-20	Structural Steel
-21	Architecturally Exposed Structural Steel
-22	Tubular Steel
-30	Structural Aluminum
-31	Architecturally Exposed Structural Aluminum
-50	Steel Wire Rope
-60	Framing Systems
-61	Space Frames
-62	Geodesic Structures
05170-05199	(Reserved)
05200	METAL JOISTS
-10	Steel Joists
-11	Standard Steel Joists
-12	Custom Fabricated Steel Joists
-20	Aluminum Joists
05230-05299	(Reserved)
05300	METAL DECKING
-10	Metal Roof Deck
-11	Steel Roof Deck
-12	Aluminum Roof Deck
-20	Metal Floor Deck
-21	Steel Floor Deck
-22	Aluminum Floor Deck
05330-05399	(Reserved)
05400	COLD-FORMED METAL FRAMING
-10	Load Bearing Metal Stud System
-20	Cold-Formed Metal Joist System
05430-05499	(Reserved)
05500	METAL FABRICATIONS
-01	Anchor Bolts
-02	Expansion Bolts
-10	Metal Stairs
-15	Ladders
-20	Handrails and Railings
-21	Pipe and Tube Railings
-30	Gratings and Floor Plates
-40	Castings
-50	Custom Enclosures
-51	Heat-Cooling Unit Enclosures
05560-05699	(Reserved)
05700	ORNAMENTAL METAL
-10	Ornamental Stairs
-15	Prefabricated Spiral Stairs
-20	Ornamental Handrails and Railings
-30	Ornamental Sheet Metal
05740-05799	(Reserved)
05800	EXPANSION CONTROL
-01	Interior Expansion Joints
-02	Exterior Expansion Joints
-05	Expansion Joint Covers
-10	Bridge Expansion Control
-11	Bridge Sole Plates
-12	Bridge Bearings
-20	Bridge Expansion Joints
05830-05899	(Reserved)
05900	METAL FINISHES
-01	Anodic Coatings
-02	Enamel Coatings
-03	Acrylic Coatings
-04	Urethane Coatings
-10	Galvanizing
05920-05999	(Reserved)

Figure 5-2 CSI Division Numbering. (The Construction Specifications Institute. CSI Document MP-2-1 MASTERFORMAT. Master List of Section Titles and Numbers. 601 Madison Street, Alexandria Virginia.

A

KEY WORD

	Broadscope Number
abattoir equipment	11500
abrasion-resistant coatings	09800
abrasive aggregates	
concrete, cast-in-place	03300
concrete, precast	03400
floor topping, heavy-duty concrete	03350
roads & walks	02500
special flooring	09700
terrazzo	09400
abrasive metal treads & nosings	05500
abrasive safety floors	09700
abrasive tile	09300
	09650
absorber plates, solar	13980
absorption boiler	15650
absorption chiller	13980
absorption refrigeration units	15650
centrifugal combination	15650
hot water activated	15650
steam activated	15650
absorption separators	15650
AC generators	16200
AC motor controls	16100
ALSO SEE SPECIFIC EQUIPMENT	
AC motors	16100
ALSO SEE SPECIFIC EQUIPMENT	
accelerators, concrete	
concrete, cast-in-place	03300
concrete, precast	03400
roads & walks	02500
access doors & hatches	
basement	08300
ceiling	08300
	09500
chutes	14500
duct	15800
equipment	SEE SPECIFIC EQUIPMENT
floor	08300
roof	07800
sidewalk	08300
wall	08300
access floor	10270
access restrictions, site	01550
access road	02500
access roads, temporary	01550
access stairs, folding	11450
accessible ceilings	09500
	13070
accessories	
acoustical	09500
bath	10800
ceramic tile	09300
chalkboard	10100
chutes	14500
commercial laundry	11110
concrete	03250
curtain	12500
drapery	12500
drywall	09250
duct	15800
electrical	16100
fireplace	10300

accessories (continued)	
formwork	03100
furnishing	12600
gypsum deck	03500
gypsum wallboard	09250
insulating roof deck	03500
interior decorating	12600
lath and plaster	09200
lightgage framing	05400
lighting	16500
open-web joists	05200
partition systems	10600
pipe	15050
plastering	09200
reinforcement	03200
roof	07800
storage shelving	10670
structural metal	05100
toilet & bath	10800
wiring	16100
yard	02440
accordion folding doors	08300
accordion folding partitions	10600
accumulator tank	15650
acid distribution systems	15300
acid etching	03350
acid-resistant brick	04200
acid-resistant coatings	09800
acid-resistant glass	08800
acid-resistant mortars	04100
acid-resistant pipe	15050
acoustic block	04200
acoustic door	08100
	08200
	08300
acoustic isolation	13080
acoustic isolation, mechanical equipment	15200
acoustic suspension systems	09100
acoustical baffles	09500
	13030
	13080
acoustical calking	07900
	09250
	09500
	10600
acoustical ceilings	09500
	09500
	13070
acoustical doors	08100
	08200
	08300
acoustical insulation	07200
	09500
	10600
	15250
	15800
acoustical masonry units	04200
acoustical metal decking	05300
acoustical panels	09500
	13070
acoustical plaster	09200
acoustical shell systems	11060
acoustical systems	09500
	13070
acoustical tile	09500
	13070
acoustical treatment	09500
ducts	15800
hangers	09500
mechanical equipment	15200

Figure 5-3 CSI Key Word Index. (The Construction Specifications Institute. CSI Document MP-2-1 MASTERFORMAT. Master List of Section Titles and Number. 601 Madison Street, Alexandria Virginia.)

UNIFORM DOCUMENTATION SYSTEM

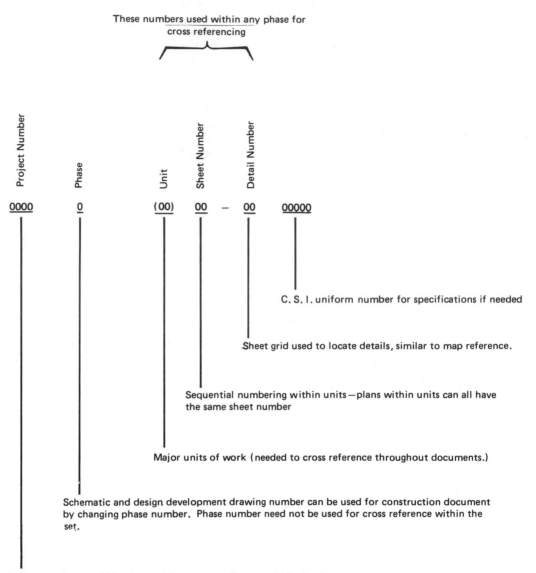

*System proposed by the Baltimore Chapter Construction Specifications Institute

Figure 5-4 Uniform Documentation System. (Baltimore Chapter Technical Reports. CSI-TRS 1972–1973. Baltimore Chapter, Construction Specifications Institute.)

5-2.2 Construction Industry Thesaurus.

The *Construction Industry Thesaurus* was developed in England to provide a basic vocabulary for the construction industry. Terminology repre-senting current construction practices are documented in the thesaurus for the purpose of storing and retrieving information. Every term in the classified display has been given a reference number. This number enables individual terms

INDEX TO DETAIL NUMBERING SYSTEM
(ACCORDING TO UNIFORM SYSTEM)

DETAILS TO BE BOUND IN SPECIFICATIONS SHALL BE NUMBERED AS INDICATED IN THIS INDEX.

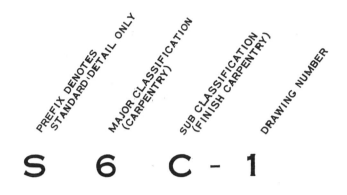

PREFIX DENOTES STANDARD DETAIL ONLY

MAJOR CLASSIFICATION (CARPENTRY)

SUB CLASSIFICATION (FINISH CARPENTRY)

DRAWING NUMBER

S 6 C - 1

NOTE: . PREFIX USED ONLY ON STANDARD DETAILS

THIS INDEX ALSO USED WITH *HTB* PRODUCT LITERATURE FILES AND MAJOR CLASSIFICATIONS USED IN SPECIFICATION, SWEETS FILE AND SHOWCASE MICRO FILM INDEX.

(SMALL NUMBERS IN INDEX USED IN FILING ONLY — NOT ON DRAWINGS.)

1

HTB, Inc., Architects • Engineers • Planners
OKLAHOMA CITY — TULSA — WASHINGTON D.C.

Figure 5-5 Index to Detail Numbering System (HTB, Inc., Architects/Engineers/Planners, Oklahoma City, Oklahoma).

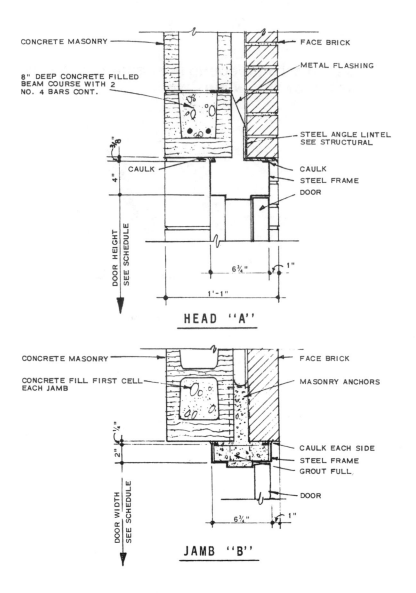

CONCRETE MASONRY

8" DEEP CONCRETE FILLED BEAM COURSE WITH 2 NO. 4 BARS CONT.

FACE BRICK

METAL FLASHING

STEEL ANGLE LINTEL SEE STRUCTURAL

CAULK

CAULK

STEEL FRAME

DOOR

DOOR HEIGHT SEE SCHEDULE

6¾"

1"

1'-1"

HEAD "A"

CONCRETE MASONRY

CONCRETE FILL FIRST CELL EACH JAMB

FACE BRICK

MASONRY ANCHORS

CAULK EACH SIDE

STEEL FRAME

GROUT FULL

DOOR

DOOR WIDTH SEE SCHEDULE

6¾"

1"

JAMB "B"

SCALE: 1½" = 1'-0"

EXTERIOR DOOR DETAILS

PROJECT NO.	DRAWING NO.
STANDARD	S8E-4

HTB, Inc., Architects • Engineers • Planners
OKLAHOMA CITY — TULSA — WASHINGTON D.C.

REMARKS:				DESCRIPTIVE TITLE		STD. SHEET NO.
MARK	REVISIONS	DATE	BY	DRAWN		
				CHECKED		
				DESIGNER		
				JOB CAPT.		
				PROJ. MGR.		
				DIRECTOR		
				ASST. DIRECTOR		

Figure 5-6 Exterior Door Details (HTB, Inc., Architects/Engineers/Planners, Oklahoma City, Oklahoma).

4 Roof Vents
H SHEET METAL WORK
  1 Ductwork; see 15 AIR-TEMPERING SYSTEM
  2 Downspouts
  3 Facias & Copings; see 5 ORNAMENTAL METAL
  4 Gutters
  5 Scuppers
  6 Sheet Metal Roofing
  7 Vermin Shields
I SHINGLES & ROOFING TILES
  1 Asbestos-Cement
  2 Asphalt
  3 Backing Material
  4 Concrete
  5 Fired Clay
  6 Metal
  7 Porcelain Enamel
  8 Slate
  9 Wood
J WALL FLASHING
  1 Expansion Joints
  2 Felts & Fabrics
  3 Metal & Metal-Coated
  4 Preformed Units
  5 Synthetic Sheets
K WATERPROOFING
  1 Hydrolithic Waterproofing
  2 Integral Waterproofing; see 3 CAST-IN-PLACE CONCRETE
  3 Liquid Waterproofing
  4 Membrane Waterproofing
  5 Metallic Oxide Waterproofing
  6 Preformed Elastic Sheets
L MISCELLANEOUS

8. DOORS, WINDOWS, & GLASS
A GENERAL INFORMATION
B CURTAINWALL SYSTEM
  1 Curtainwall Panels
  2 Mullion Systems
  3 Panel Systems
  4 Precast Concrete Panels; see 3 PRECAST CONCRETE
C FINISH HARDWARE
  1 Bolts
  2 Cabinet Hardware
  3 Casement Openers
  4 Closers & Checks
  5 Doorstops & Holders
  6 Exit Devices
  7 Hinges
  8 Key Cabinets
  9 Kick & Mop Plates
  10 Locksets & Latchsets
  11 Push & Pull Units
  12 Sash Balances
  13 Sash Cleaners' Hooks
  14 Sash Latches & Lifts
  15 Special Knobs & Trim
  16 Thresholds
D GLASS & GLAZING
  1 Coatings for Glass
  2 Corrugated Glass
  3 Decorative & Obscure

4 Glass Masonry Units; see 4 UNIT MASONRY
  5 Glazing Sealants; see 7 CALKING & SEALANTS
  6 Insulating Glass
  7 Laminated Glass
  8 Lead Cames; see 12 ARTWORK
  9 Lead Glass
  10 Mirrors
  11 Opaque Glass
  12 Plate Glass
  13 Sheet Glass
  14 Solar-Controlling
  15 Stained Glass Work; see 12 ARTWORK
  16 Tempered Glass
  17 Translucent Plastics
  18 Wired Glass
E METAL DOORS & FRAMES
  1 Custom
  2 Fire-Rated
  3 Flush Non-Ferrous
  4 Flush Steel
  5 Louvered
  6 Paneled Non-Ferrous
  7 Paneled Steel
  8 Tabular Non-Ferrous
  9 Tubular Steel
F METAL WINDOWS
  1 Aluminum
  2 Bronze
  3 Stainless Steel
  4 Steel
G OPERATORS
  1 Doors
  2 Windows
H SPECIAL DOORS
  1 Blast-Resistant
  2 Coiling
  3 Elevator Doors; see 14 ELEVATORS
  4 Flexible
  5 Folding
  6 Hangar
  7 Hoistway Doors; see 14 ELEVATORS
  8 Insulated Doors; see 13 INSULATED ROOMS
  9 Jalousie
  10 Kalamein
  11 Overhead, Sectional
  12 Overhead, Swing-Up
  13 Packaged Units
  14 Radiation-Retarding
  15 Revolving
  16 Rolling
  17 Screen & Storm
  18 Sliding Glazed
  19 Sound-Retarding
  20 Tempered Glass
  21 Vault Doors; see 13 STORAGE VAULTS
  22 Vertically Sliding
I STOREFRONT SYSTEM
  1 Entrances
  2 Facings
  3 Framing Systems
  4 Moldings & Trim
  5 Operable Sash; see 8 METAL WINDOWS

5

HTB, Inc., Architects · Engineers · Planners
OKLAHOMA CITY — TULSA — WASHINGTON D.C.

REMARKS:				DESCRIPTIVE TITLE		STD. SHEET NO.
MARK	REVISIONS	DATE	BY	DRAWN		
				CHECKED		
				DESIGNER		
				JOB CAPT.		
				PROJ. MGR.		
				DIRECTOR		
				ASST. DIRECTOR		

Figure 5-7 Specification Division Indexing (HTB, Inc., Architects/Engineers/Planners, Oklahoma City, Oklahoma).

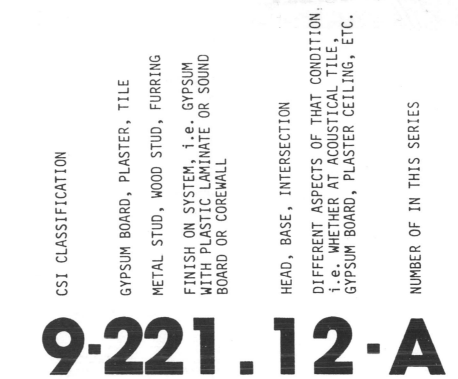

CSI CLASSIFICATION

GYPSUM BOARD, PLASTER, TILE

METAL STUD, WOOD STUD, FURRING

FINISH ON SYSTEM, i.e. GYPSUM WITH PLASTIC LAMINATE OR SOUND BOARD OR COREWALL

HEAD, BASE, INTERSECTION

DIFFERENT ASPECTS OF THAT CONDITION, i.e. WHETHER AT ACOUSTICAL TILE, GYPSUM BOARD, PLASTER CEILING, ETC.

NUMBER OF IN THIS SERIES

9-221.12-A

9-221.12	Gypsum board partition at acoustic panel ceiling
9-221.13	Gypsum board partition at gypsum board ceiling
9-221.22	Gypsum board partition base at concrete
9-221.23	Gypsum board partition base at ceramic tile
9-225.12	Gypsum board with sound board partition at acoustic panel ceiling
9-225.13	Gypsum board with sound board partition at gypsum board ceiling
9-225.22	Gypsum board with sound board partition base at concrete
9-241.12	Furred gypsum board at acoustic panel ceiling
9-241.22	Furred gypsum board base at concrete
9-321.12	Plaster metal stud partition at acoustic panel ceiling

Figure 5-8 Detail Identification by CSI Classification (Clark & Van Voorhis Architects, Inc., Phoenix, Arizona).

within the classified display to be located by reference to the index. All terms prepared for the thesaurus are listed alphabetically to:

1. indicate synonyms
2. indicate the location of code terms in the classified display
3. show relationships between terms.

The documentation of each term used in the thesaurus has an overall conceptual relationship to the general body of information generated in the construction industry. Ten major facets of information have been recognized in the documentation of all terms identified in the construction thesaurus. The following list of categories identifies each facet of information.

Classification of Subjects in Construction Industry Thesaurus

A. Forms of Records
B. Peripheral Subjects
C. Time
D. Place
E. Properties and Measures
F. Agents (of construction)
G. Operations and Processes
H. Materials
J. Parts of Construction Works
K. Construction Works

Terms identified in each of the Ten facets form the basis of the classified display. Eight core facets, C through K, are recognized as most appropriate for the construction industry. The following is an example of two facets:

MATERIALS

▷ *Scope.* Materials used in construction (the constituent material from which, for example, doors, beams, or nails are made).

▷ *Examples.* Concrete, cement, glass, steel, wood.

PARTS OF CONSTRUCTION WORKS

▷ *Scope.* All the physical parts of buildings and other works.

▷ *Examples.* Roofs, walls, doors, beams, window frames, nails.

Within each facet there are divisions and subdivisions of the broader term. Terms are selected and positioned by their characteristics. An example of the characteristic relationship between terms is demonstrated in the expansion of the term "walls" shown in Figure 5-9.

A numbering system is developed for all terms identified within the ten facets. The first letter in the alphanumeric system denotes the facet in which the term is displayed. Following the letter are five numeric digits identifying a sequential order [14]. To modify this system for detail banking requires a thorough review of all terms and expansion of the numbering system. The following procedures should be taken to utilize this system.

Detail Filing Using Construction Industry Thesaurus (England)

1. Each detail is filed under term selected from classification J = parts of construction works.
2. Details are given numbers associated with selected descriptor terms.
   Example. A. Wall details are given a J96010 number.
   B. Internal wall details are given a J98010 number.
   C. If more than one detail is collected in a descriptor category, a three- or four-digit number can be added: J96010-0001.
3. Details can be filed numerically in notebooks organized for each facet of information.
4. Terms and numbers associated with details can also be stored in word processors and computers.
5. Option = select appropriate descriptor terms

CIT
CLASSIFICATION
NUMBERING

Functional parts of construction works
Parts for division, structure and circulation
Dividing elements
Horizontal dividing elements

J96010	Walls = Vertical dividing elements	
J94210	Friezes	*Parts of walls by position*
J96410	Aprons	*ditto*
J96415	Fascia boards	*ditto*
	Note: All kinds of walls are given below for convenience. In indexing systems using compound headings Walls which serve a structural function should be indexed as Structure/Walls.	
J96510	Non loadbearing walls	*Walls by "loadbearing" characteristic*
J96520	Loadbearing walls	
	Note: Index as Structure/Walls	
J96530	Bearing walls	
J97010	External walls	*Walls by position*
J97210	Gables = Gable ends	*External walls by position*
J97410	Parapets	*ditto*
J97610	Curtain walls	*External walls by support*
J98010	Internal walls = Dividers = Space dividers	*Walls by position*
J98410	Party walls	*Walls by ownership*
J98510	Brick walls	*Walls by components*
J98610	Cavity walls	*Walls by construction*
J98620	Screen walls	*Walls by function*
J98625	Partitions	*ditto*
J98630	Rood screens	*ditto*
J98635	Room dividers	*ditto*
J98640	Boundary walls	*ditto*

Figure 5-9 Example of Construction Industry Thesaurus Classification and Numbering (*Construction Industry Thesaurus.* Property Services Agency, Department of the Environment, England).

from CIT and apply procedures identified in Chapter 6.

5-2.3 Alphanumeric systems emerging in private firms.

Private firms are beginning to develop independent information-handling systems for special in-house data filing. These systems are emerging as a result of the limited information-handling programs available for design professionals. The lack of guidance for developing information-handling programs is causing private firms to create their own unique systems.

Most emerging systems are a combination of alphanumeric and special category numbering associated with different classifications of information. At the outset these systems seem to satisfy all the information-handling needs of the firm. However, as the body of information grows, these systems become more complex and difficult to use in conducting specific searches. In many cases the system must be amended in order to add new types of information. The example that follows demonstrates the alphanumeric framework generated for these new systems.

Indexing Information Using An Alphanumeric System

1. Categories and classifications are developed for different types of information.
   Example. A. *Categories*
   Roof systems = RS.
   Exterior wall systems = EW.
   Interior wall systems = IW.
   B. *Classifications—Material*
   Wood = 1.
   Concrete = 2.
   Steel = 3.
   C. Subject areas for categories and classifications vary for different private systems. Building types, construction disciplines, and areas of design are also used to structure alphanumeric systems.

2. A numbering system is required to group details into categories.
   Example. A. Wood roof system details are placed in an RS category and classified number 1 = RS-1.
   B. If more than one detail is collected in a category, a three- or four-digit number is added: RS-1-0001.

3. Details are filed numerically by categories in notebooks or catalogs.

The following three examples of filing systems show individual organization efforts and experiences in working with information storage and retrieval systems. Minor variations in these systems further point the need to coordinate research and experimental efforts so that a "master" or "universal" system can be developed for design professionals.

Example 1

Architectural Small Sheet Drawings*

Filing System

Categories	Standard	Special
General (use sparingly)	100A–199Z	1000–1999
Foundations	200A–299Z	2000–2999
Floors	300A–399Z	3000–3999
Walls (exterior)	400A–499Z	4000–4999
Roofs	500A–599Z	5000–5999
Soffits (exterior)	600A–699Z	6000–6999
Partitions (interior)	700A–799Z	7000–7999
Ceilings	800A–899Z	8000–8999
Grade improvements/ landscaping	900A–999Z	9000–9999

*Source: Thorsen & Thorshov Architects, Inc., Minneapolis, Minnesota.

Each category represents a major architectural surface. The sequence of categories is similar to a general sequence of construction for buildings. To search for the existence of a previously drawn detail (or other small sheet drawing), select the appropriate category from the surfaces listed. Select a detailed surface or one that relates closest to the detail (e.g., by penetration through the surface or attachment of equipment to the surface).

For a drawing indicating two surfaces (e.g., floor edge at a wall), select the category that corresponds to the predominant problem being solved for the contractor. If both surfaces are of equal importance, select the surface that will be built first. Where one is still unable to decide which category is best, the drawing can be filed under all categories that apply.

If the drawing is for a typical condition that may occur in more than one building, it should be classified in the "standard" hundred series that corresponds to its category. Each number within a "standard" series indicates the detail's building location/function. Alphabetical letter suffixes indicate a detail's major component or type of system construction.

If the drawing is for only one building or very limited application, it should be classified in its respective thousand series. These numbers are assigned in consecutive sequence as they are drawn. They do not have suffix letters. The numbers for the "special" series do not indicate any particular function.

This in-house filing system has been developed because the original specification numbering format did not adequately define a single category when more than one product or material is indicated in a drawing. Therefore it did not permit direct retrieval or access to details, that is, without referring to a previous project, nor did it provide a readily apparent categorization of details for filing. This system has attempted to overcome just such deficiencies. For details that were originally categorized according to the specification format, an index has been provided with both the old specification numbers and the updated file numbers for older details that are now classified.

☐

Example 2

Retrievable Detail Index (REDI) System*

Letters are used for the various categories of details. These letters head sections of the index are used in numbering the drawings within the manual. The drawing letter in the circle would then be cross-referenced to the standard working drawing sheets.

▷ Ⓐ MASTER MANUAL

▷ Ⓑ BUILDING CODE & ZONING DATA BLANKS/DRAWING NUMBERS

▷ Ⓒ CURTAINWALL/WINDOWS/ ENTRANCES/STOREFRONT

▷ Ⓓ DOOR & FRAME DETAILS & ELEVATIONS

▷ Ⓔ EXTERIOR DETAILS

▷ Ⓕ FINISH SCHEDULE/DOOR SCHEDULE

▷ Ⓖ GRAPHIC SYMBOLS/ ABBREVIATION

▷ Ⓗ HANDRAILS/STAIRS/RAMPS

▷ Ⓘ INTERIOR DETAILS

▷ Ⓜ MILLWORK

▷ Ⓟ PARTITION-TYPE SECTIONS

*Source: Smith Entzeroth, Inc., Architects/Planners, Clayton, Missouri.

▷ Ⓢ SITEWORK DETAILS

▷ Ⓣ TOILET DETAILS & MOUNTING HEIGHTS

Example.

Ⓓ DOOR & FRAME DETAILS & ELEVATIONS

1. Typical hollow metal frame sections.
2. Typical hollow metal and wood view panel.
3. Door louver detail.
4. Door and frame elevation sheets.
5. Hollow metal borrowlite frame.
6. Wood door frame and sidelight frame.
7. Overhead door and track.
8. Roll-up grille and door.
9. Pocket door detail.
10. Door thresholds.
11. Penthouse door sills.
12. Elevator doors.
13. Masonry block door lintel types.

□

Example 3

Drawing Format System*

Drawing Format System

Drawing types series 000—general through series 400—$\frac{1}{4}$ scale plans and elevations will be drawn on office standard sheets 30 × 42 with printed title blocks. Series 500—details, exterior through series 900—details, interior will be drawn on office standard sheets $8\frac{1}{2}$ × 11 and will be issued either on $8\frac{1}{2}$ × 11 or pasted up on large sheets, a wash-off mylar made and issued as part of the large sheet set.

*Source: Jung/Brannen Associates, Boston, Massachusetts.

Drawing Series Breakdown—10's Digit

For all projects the large-size sheets will be divided into flexible breakdowns and each of which will be assigned by the Project Architect. The following is a sample breakdown but the actual list will be left to the discretion of the Project Architect:

000	GENERAL
000	Title sheet and drawing index
010	General notes
020	Abbreviations, material designations (if not in $8\frac{1}{2}$ × 11)
030	Drawing system explanation (if not on $8\frac{1}{2}$ × 11)
100	$\frac{1}{8}$"-SCALE PLANS
110	Lowest-level plan
120	Next lowest-level plan
130	etc.

Note: The units number would be the type floor plan.

111	floor plan: dimensions
112	floor plan: notes, symbols, wall keys
113	reflected ceiling plan
200	$\frac{1}{8}$" EXT ELEVATIONS & BUILDING SECTIONS
210	Exterior elevations (211–219)
220	Building sections (221–229)
230	Interior elevations (231–239)
300	$\frac{1}{2}$" WALL SECTIONS
300	Consecutive (if appropriate for project, they could be broken down into convenient sections: 310—Exterior, 320 Interior, etc.)
400	LARGE-SCALE PLANS & ELEVATIONS
400	Schedules
410	Toilet plans
420	Toilet elevations
430	Interior plans
440	Interior elevations
450	Stair elevations and sections
460	Stair elevations and sections

470   Elevator plan
480   Elevator sections

Details

All details will be drawn on $8\frac{1}{2} \times 11$ standard preprinted sheets, using the basic box sizes as indicated by the numbering system for the small-size sheets. The letter or letters between the number and the type detail listed here is the abbreviation to be used as part of the individual detail number in the box at the lower left corner of the detail. The following breakdown will always be used:

500   Details, Exterior
600   Details, Doors
700   Details, Windows/louvers
800   Details, Finishes
900   Details, Interior

Breakdown of Details

500   EXTERIOR
500   N   Notes and criteria
510   SC   Schedules
520   F   Below grade (foundation)
530   S   Site and paving
540   W   Wall
550   SF   Soffit and fascia
560   P   Parapet
570   R   Roof
580   EX   Expansion control
590   MS   Miscellaneous exterior

600   DOORS
600   N   Notes and criteria
610   SC   Schedules
620   E   Door type (elevation)
630   F   Frame type (elevation)
640   HM   Hollow metal sections details
650   ST   Steel sections and details

660   WO   Wood sections and details
670   AL   Aluminum sections and details
680   OH   Overhead/rolling sections and details
690   SP   Special sections and details

700   WINDOWS/LOUVERS
700   N   Notes and criteria
710   SC   Schedules
720   E   Window and louver types (elevations)
730   HM   Hollow metal details
740   ST   Steel details
750   WD   Wood details
760   AL   Aluminum details
770   SP   Special details

800   FINISHES
800   N   Notes and criteria
810   SC   Finish schedule
820   WC   Wall core schedule and details
830   WF   Wall finish schedule and details
840   F   Floor schedule and details
850   CL   Ceiling schedule and details
860   B   Base schedule and details
870   WC   Wall to ceiling schedule and details
880   WW   Wall to wall schedule and details
890   R   Rated/assemblies

900   INTERIOR
900   N   Notes and criteria
910   SC   Schedules
920   ST   Stairs
930   EL   Elevator
940   MM   Miscellaneous metal
950   T   Toilet and locker room—CSI Div 10
960   M   Millwork
970   EX   Expansion control

980 SP   Special interior
990 ID   Identifying devices

Details

For small-size sheets or details pasted up on large-size sheets, the drawing identification number will be as follows:

1. *8½" × 11" Sheets.* The number in the lower right-hand corner on small-size sheets shall consist of drawing type, the drawing-type breakdown; the units number shall always be zero, followed by a dash and then the sheet number starting with one and running consecutively to as high as required. The discipline letter and expanded number shall be dropped.

Example:

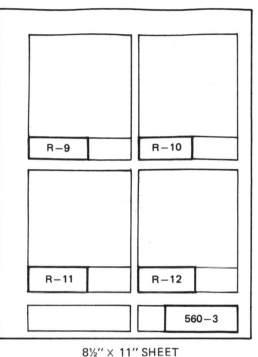

Example:

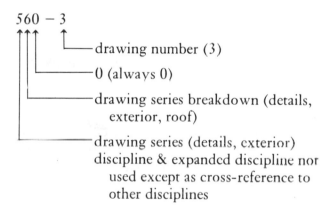

560 − 3
- drawing number (3)
- 0 (always 0)
- drawing series breakdown (details, exterior, roof)
- drawing series (details, exterior) discipline & expanded discipline not used except as cross-reference to other disciplines

The detail(s) on the sheet will be numbered by the abbreviations listed in the office detail filing system (See Mandatory Conventions).

8½" × 11" SHEET

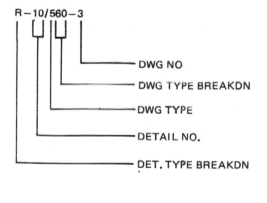

R−10/560−3
- DWG NO
- DWG TYPE BREAKDN
- DWG TYPE
- DETAIL NO.
- DET. TYPE BREAKDN

RO−10

560−3

2. *Large Sheets.* If the details are copied and pasted up on a large size sheet and made into a wash off mylar, then the system used to identify the large size sheets will be used except that in place of a letter designation for the elements on the sheet, the detail abbreviations as found in the detail filing system will be used, followed by the detail number.

Example:

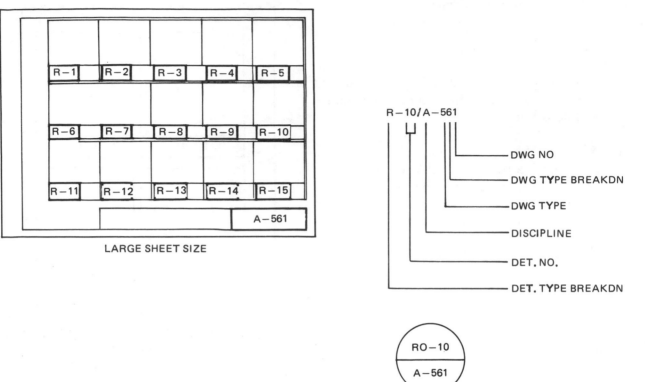

LARGE SHEET SIZE

5-3 HOW TO EVALUATE A SYSTEM

Some indexing systems have been keyed into construction specifications. Others have developed separate project manuals or a unique system for filing construction details. No matter what system is selected, it must be consistent and relate to the overall information-handling plan. Systems that have been coordinated with specifications can be somewhat limited because it is difficult to identify the section in which the construction detail should be stored. There may be great overlap with other divisions, therefore, the individual preparing the specification must make the choice as to in which section the data should be stored. A number of alternatives complicates the system.

Subject areas formed into categories can make it easier for users to enter the system and pursue the selection of a descriptor for retrieving a specific construction detail. If a classification and category system is used, it should be related to the building or structural system for which construc-

tion details are being developed. For example, the building envelope can be classified into different categories such as roofing systems, wall systems, and foundation systems. These categories can contain a number of specific subject areas for the appropriate details. The search process can be carried out by first selecting the appropriate category and then a subject descriptor relating to the specific construction detail.

Each system has its unique characteristics which can be evaluated for overall effectiveness and performance. When one evaluates a system, it is critical to keep in mind the following attributes:

▷ Scope
▷ Scale
▷ Time
▷ Space
▷ Organization
▷ Relationships
▷ Control

▷ Directness
▷ Flexibility
▷ Understandability

It is difficult to identify all of the peculiar and distinct qualities associated with each system. However, user feedback, surveys, and observations have produced the following advantages and disadvantages that can guide the system evaluator.

5-3.1 Indexing by specification systems.

Advantages

▷ Relates to architectural/engineering subject areas.
▷ Works well for material and product catalogs.
▷ Numbering and categories accepted by a national organization.
▷ Draftspersons and specifications writers can interact on detail and material selections.
▷ Is rapidly increasing in overall acceptance by the construction industry.

Disadvantages

▷ Limited to subjects and categories found in specification divisions.
▷ Subjects and numbers can be changed by agreement of membership in organization.
▷ Understood more by specifications writers than by draftspersons, project architects, and engineers.
▷ System not being used by all segments of the construction industry.
▷ Overlap with other divisions causes indexers indecision in filing information.
▷ Details consisting of more than one material make it difficult to select the best division or section.

5-3.2 Alphanumeric systems.

Advantages

▷ Originator of system is free to select and arrange letters and numbers as determined by in-house data requirements.
▷ System can be developed over a short time frame.
▷ System is independent of outside controls.

Disadvantages

▷ Systems are generally limited to smaller-scale data-collection programs.
▷ Categories are usually too general and too broad in scope for a specific search.
▷ Indexer is forced to decide which category detail is best filed.
▷ Searcher may not use the system as indexer intended.
▷ Relationships between categories and classifications make it difficult to conduct a specific detail search.
▷ Interrelationship of categories and classifications can lead to confusion and misunderstanding.
▷ Letters and numbers can become confusing as system expands.
▷ Individual systems tend to require more user training and special guidance.
▷ Potential for system success can be reduced when originator of program leaves organization.

5-3.3 Dewey Decimal and Library of Congress classification systems.

Disadvantages

▷ Consumes considerable staff time, and architects/engineers are not generally familiar with their use.

5-3.4 The descriptor term system is the recommended system.

Advantages

▷ Terminology can be generated that is acceptable and standard within the construction industry.

▷ New terms and descriptors can be added that reflect changes in technology and methods of construction.

▷ Family relationships of terms can be defined to group-related areas of information.

▷ Narrow terms, used for terms, and related terms, can be used to achieve cross-referencing of information.

▷ All members of the construction industry have an opportunity to agree on an acceptable terminology.

▷ Subject areas become the key entry to the system rather than symbols, abbreviations, and other abstract methods of indexing.

▷ Multidescriptor indexing increases potential to obtain specific items of information.

▷ Have existing thesauri to use as reference standards in developing acceptable terminology.

Disadvantages

▷ Need to develop a thesaurus of acceptable terminology.

▷ Requires agreement on selection of acceptable and standard terminology.

SIX
DETAIL BANKING PROCEDURES: HOW TO DEVELOP A MASTER SYSTEM

6-1 HOW TO DEVELOP THE DESCRIPTOR TERM SYSTEM

Research and experience gained in information handling has provided sufficient evidence to support further development of what is known as the "descriptor term system" [8]. The descriptor term or subject area word selection represents key elements of each detail. *The construction language and thesaurus discussed earlier in Chapter 4 consists of a collection of these word descriptors.* Example. Descriptor term = *Insulation.*

A descriptor or a subject area word selection represents an agreed-upon language that is acceptable within the construction industry. Thus the descriptor term system is simple, direct, and based on the overall language of a specific discipline area. Only three major steps must be fulfilled:

1. The user must develop a construction language and thesaurus. (Note: This step should not be viewed as a major undertaking since descriptor terms selected for detail banking automatically form the construction language. Most design organizations already have a large percent of acceptable construction terms in use for working drawings and specifications.)

2. The user must identify acceptable descriptor terms for indexing details.

3. The user must create a numbering system to file collected details.

The three types of thesauri cited in Chapter 4 should be analyzed for potential use:

1. *Construction Industry Thesaurus* in England [14].

2. *Thesaurus of ERIC Descriptors* developed for the U.S. Department of Health, Education, and Welfare [13], [15].

3. *Canadian Thesaurus of Construction Science and Technology* developed in the early 1970s for use in a potential construction information system, and revised in 1978 [11].

These thesauri provide an excellent base for establishing a national or in-house information-handling program. Subject area descriptors should be selected using the guidelines established by the Joint Council of Engineers [12]. These guidelines can aid design professionals in selecting appropriate terminology for generating a construction language. Additional guidelines for selecting terms are cited in step 5 at the beginning of Section 6-2. Selected guidelines taken from *Rules for Thesaurus Preparation, ERIC 1969* are as follows [9]:

1.1
Descriptor Elements

1.1.1
Descriptor Selection
A descriptor is any single or multiword term that appears in the thesaurus and that may be

used for indexing a document. Rules for selecting descriptors follow.

1.1.1.1
Descriptors should represent important concepts found in the literature rather than concepts derived independently. They should also reflect the language used in the literature to describe such concepts.

1.1.1.2
Descriptors selected should have an agreed-upon meaning by relevant user groups and should be acceptable terminology for that user group. Acceptability will involve decisions as to obsolescence, negative connotations, colloquial usage, and other factors. (See also 1.1.3.)

1.1.1.3
Since frequency of occurrence of terms is a factor in establishing descriptors, records should be kept of the number of times a candidate term has been used in indexing and/or searching.

1.1.1.4
Multiword descriptors (bound terms, precoordinated terms, and others) should be used whenever single-word descriptors cannot describe a concept adequately or provide effective retrieval. Many problems of this type can be solved by the careful application of rule 1.1.1.1. The following points should also be considered.

1.1.1.4.1
Use of a multiword descriptor is justified if any of the individual words in the multiword descriptor can combine so frequently with other descriptors as to produce many false coordinations.

1.1.1.4.2
Use of a multiword descriptor to represent a unique concept is justified if the individual words of that multiword term are also unique descriptors that, when coordinated with each other, represent concepts different from the one intended by the multiword term.

Example.
STUDENTS
TEACHERS
STUDENT TEACHERS

1.1.1.4.3
If a single-word term (used as a substantive) is so general as to be virtually useless in searching (e.g., SCHOOLS), consider the use of that term with another term (e.g., SECONDARY SCHOOLS).

1.1.1.4.4
Multiword descriptors, like single-word descriptors, must be carefully considered for placement in descriptor hierarchies.

1.1.1.4.5
Do not use inverted entries. (See Section 4.0.)
Examples.
DEVELOPMENT, EMOTIONAL
EDUCATION, ADULT are not valid descriptors; EMOTIONAL
DEVELOPMENT ADULT
EDUCATION are valid descriptors.

1.1.2
Cross-References

1.1.2.1
Use (USE)
The "USE" reference is utilized in two situations:

1.1.2.1.1
To indicate preferred usage.
Example.
Advanced Education, USE HIGHER
EDUCATION

1.1.2.1.2
To cross-reference an abbreviation.
Example.
ETV, USE EDUCATIONAL
TELEVISION

Note: Inverted entries are not to be included in cross-referencing. For example, SCHOOL BUILDINGS would not be cross-referenced with BUILDINGS, SCHOOL.

1.1.2.2

Used For (UF) The mandatory reciprocal of the "USE" reference is the UF reference.

Example.

> HIGHER EDUCATION, UF Advanced Education, EDUCATIONAL TELEVISION, UF ETV

1.1.2.3

Broad Term (BT) and Narrower Term (NT). The broader term (BT) and narrower term (NT) cross-references are employed to indicate any class relationships that may exist among descriptors. Where descriptors represent concepts that are included within the class of concepts represented by another descriptor, this relationship is shown by the broader term reference.

Examples.

> BUILDING DESIGN, BT DESIGN, INTERIOR, DESIGN, BT DESIGN

Each broad term reference requires a corresponding reciprocal narrower term reference:

Example.

> DESIGN, NT, BUILDING DESIGN, INTERIOR DESIGN

(*Note:* A given descriptor may be a member of more than one class. If the descriptor is a member of more than one class, reciprocal references must be made to show these relationships to the next given level of the affected classes.)

Example.

> STUDENT TEACHERS, BT STUDENTS, TEACHERS

1.1.2.4

Related Term (RT)

The related term (RT) cross reference is employed as a guide from a given descriptor to other descriptors that are closely related conceptually but that do not possess class relationships as in 1.1.2.3. In general, the RT cross-reference is given for the convenience of the user who, in examining one descriptor, needs to be reminded or informed of the existence of a related descriptor. A reciprocal relationship should always be listed.

1.1.3

Ambiguity

Parenthetical qualifiers and scope notes should be used when the intended usage of a descriptor has not been made explicit by cross-reference.

1.1.3.1

Parenthetical Qualifiers

A parenthetical qualifier identifies any particular indexable meaning of a homograph. This meaning is given the status of a descriptor by listing the homograph and its parenthetical qualifier in the thesaurus display. The homograph and its parenthetical qualifier are considered inseparable in indexing or searching. For any given homograph, there may exist as many descriptors consisting of homograph plus parenthetical qualifier as there are unique indexable meanings for that homograph. If a descriptor is judged to be a homograph, it must have a parenthetical qualifier.

Examples.

> GRADE (INCLINE), GRADES (SCHOLASTIC), POSITION (LOCATION), POSITION (TITLE)

Do not use another homograph as a parenthetical qualifier.

Example.

> Do not use GRADES (MARK). But rather, GRADES (SCHOLASTIC)

One of the reasons for restricting the use of parenthetical qualifiers to homographs is to preclude the user of inverted entries.

1.1.3.2

Scope Notes (SN)

A scope note is a brief statement appearing in the thesaurus of the intended usage of a descriptor. It need not provide a formal definition. Scope notes are generally used in two situations:

1.1.3.2.1

To restrict the usage of a descriptor.

Example.

> ARCHITECTURAL BARRIERS
> SN   Building elements that become obstacles to physically handicapped persons.

1.1.3.2.2

To explain the usage of a descriptor that may have different meanings in several areas of education.

Example.

ARCHITECTURAL PROGRAMMING

SN   The process of identification and systematic organization of the functional, architectural, structural, mechanical, and esthetic criteria which influence decision making for the design of a functional space, building, or facility.

□

6-1.1   Descriptor structure.

The following example is a typical thesaurus term structure and the required format:

Descriptor.   DESIGN NEEDS.

Scope Note.   SN   Human requirements supplied or relieved by the physical environment.

Used For.   UF   Design requirements.

Narrower Term.   NT   Physical design needs.
Psychological design needs.

Broader Term.   BT   Needs.

Related Terms.   RT   Design preferences.
Architectural programming.
Building design.

6-1.2   How to select descriptors.

Descriptors should represent agreed-upon language of the construction industry. The term "chimney" can be found in most construction thesauri. If a chimney detail was selected for the detail bank, the descriptor "chimney" could be utilized as one of the indexing terms. Additional descriptors can be assigned to the drawing if required to identify subject area coverage. For example, "chimney cap" could be a narrow term descriptor for this detail. The example merges the expanded descriptor terminology associated with *Thesaurus of ERIC Descriptors* and *Canadian Thesaurus of Construction Science and Technology* [13], [11]. An example of a construction detail and a related descriptor term are shown in Figures 6-1 and 6-2.

6-1.3   Steps to be followed in selecting descriptors.

1.   Review the detail as a potential user.

2.   Determine major goals of the detail.

3.   Select key terms that will identify the intent and purpose of the detail:

  a.   Major category of construction

  b.   Materials

  c.   Uses

  d.   Location

  e.   Building type

  f.   Object or element

4.   Identify terms that will relate the goals of the detail in a complete overview manner. Figure 6-3 demonstrates the use of primary and secondary descriptors.

5.   Review the terms abstracted from the detail and analyze each term for possible use as a major descriptor.

6.   Major descriptor terms should form a basic outline of the detail content.

7.   Review existing terms in the construction thesaurus for possible selection.

8.   Select new descriptor terms when required to satisfy the goals of the detail.

9.   Obtain user group review before new descriptor terms are added to construction thesaurus. (Note: These procedures are based on steps prepared by the author for establishing ERIC descriptors for the Clearing House on Educational Facilities.)

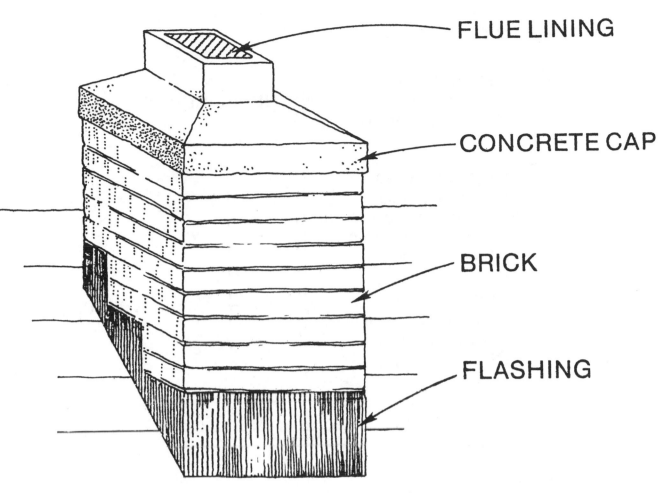

FLUE LINING

CONCRETE CAP

BRICK

FLASHING

Figure 6-1 Chimney Detail.

6-2 CRITICAL STEPS IN DETAIL BANKING

Design firms contemplating the development of a detail banking system should employ the following steps [21]:

1. Select and prepare detail for banking system.
2. Assign detail a drawing or accession number.
3. Study detail and select appropriate subject area descriptors for indexing.
4. Select terminology from existing thesauri and/or generate new terms consistent and acceptable with construction industry language.
5. If new terms must be generated, select them in accordance with:

   a. *Rules for Preparing and Updating Engineering Thesauri*, prepared by the Engineers Joint Council [12].

   b. *Manual for Building A Technical Thesaurus,* prepared by Project LEX of the Office of Naval Research [20].

   c. *ERIC Rules for Thesaurus Preparation,* prepared by the Office of Education Panel on Education Terminology [9].

6. Develop a mini-construction thesaurus based on the terminology generated by the details filed in the system. Ultimately, a national construction thesaurus should be developed jointly by all disciplines within the construction industry.

CHIMNEY

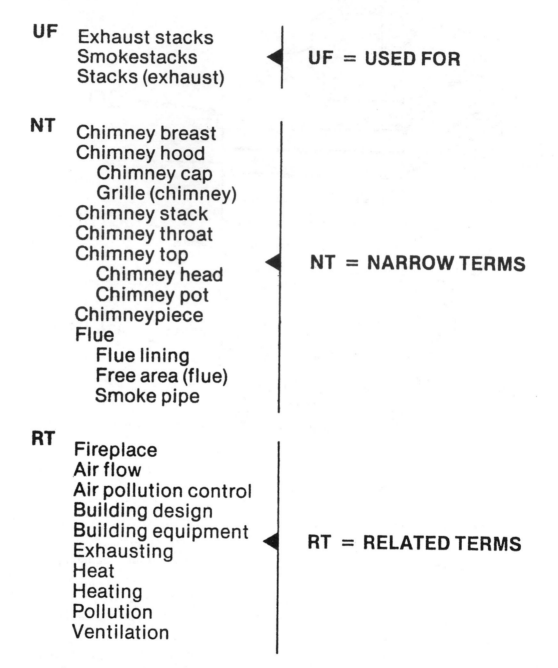

UF Exhaust stacks
Smokestacks
Stacks (exhaust)

UF = USED FOR

NT Chimney breast
Chimney hood
   Chimney cap
   Grille (chimney)
Chimney stack
Chimney throat
Chimney top
   Chimney head
   Chimney pot
Chimneypiece
Flue
   Flue lining
   Free area (flue)
   Smoke pipe

NT = NARROW TERMS

RT Fireplace
Air flow
Air pollution control
Building design
Building equipment
Exhausting
Heat
Heating
Pollution
Ventilation

RT = RELATED TERMS

Figure 6-2 Subject Area Descriptor Selection and Development.

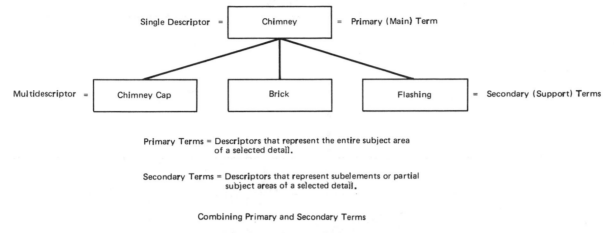

Single Descriptor = Chimney = Primary (Main) Term

Multidescriptor = Chimney Cap   Brick   Flashing = Secondary (Support) Terms

Primary Terms = Descriptors that represent the entire subject area
of a selected detail.

Secondary Terms = Descriptors that represent subelements or partial
subject areas of a selected detail.

Combining Primary and Secondary Terms

Primary terms identification should be the first level of descriptor
selection. Secondary terms should be added as required to identify
critical subelements that further define the scope of a detail.

Figure 6-3   Single Versus Multidescriptor Selection.

7. Develop a detail storage and retrieval system by preparing and filing *data cards* under appropriate descriptor terms.

8. File details consecutively by drawing number.

9. Select and develop a manual or computer system to store, search, and retrieve details. The use of word processors and computers can greatly improve the efficiency and effectiveness of the system.

APPLICATIONS

The following parapet wall and stair detail have been selected to demonstrate the processing procedures used in detail banking.

Example 1

▷ Descriptor Selection for Detail Number 000 316

▷ Parapet Wall Detail—Figure 6-4

▷ Selecting Descriptors

Step 1.   Study detail and identify critical areas of information being presented. See Section 6–1.3 for procedures.

Step 2.   Select appropriate descriptor terms for indexing detail—Figure 6-6 (page 124).

    a.   Search existing thesauri for potential descriptors.

    b.   Generate new descriptors if required to define subject area of detail.

Step 3.   Record descriptor terms for storage and retrieval on Detail Data Card and detail—Figure 6-5 (page 123).

Results

▷ The descriptors that best describe the subject of detail are "Parapet Wall" and "Masonry Construction."

▷ A search of the *Canadian Thesaurus of Construction Science and Technology* produces the following descriptors [11]:

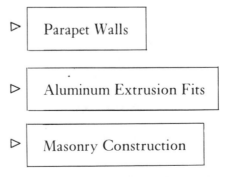

▷ Parapet Walls

▷ Aluminum Extrusion Fits

▷ Masonry Construction

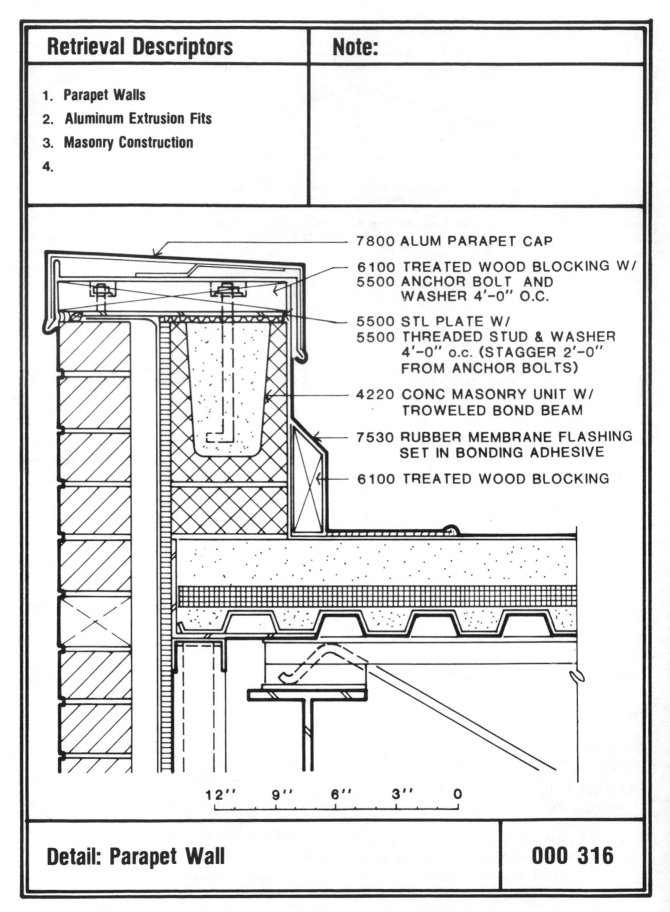

Retrieval Descriptors	Note:
1. Parapet Walls 2. Aluminum Extrusion Fits 3. Masonry Construction 4.	

Detail: Parapet Wall   **000 316**

7800 ALUM PARAPET CAP

6100 TREATED WOOD BLOCKING W/
5500 ANCHOR BOLT AND
WASHER 4'-0" O.C.

5500 STL PLATE W/
5500 THREADED STUD & WASHER
4'-0" o.c. (STAGGER 2'-0"
FROM ANCHOR BOLTS)

4220 CONC MASONRY UNIT W/
TROWELED BOND BEAM

7530 RUBBER MEMBRANE FLASHING
SET IN BONDING ADHESIVE

6100 TREATED WOOD BLOCKING

12"   9"   6"   3"   0

Figure 6-4 Retrieval Descriptors Identifying Parapet Wall Detail. (Detail Content: Gresham, Smith and Partners, Nashville, Tennessee.)

DETAIL DATA CARD

Bank Detail No.:		*000 316*
Detail Title/Description:	*PARAPET WALL*	

Scale: *AS SHOWN*	Interior	Exterior *✓*

Development Data

Developed By: *JOHN P. JONES*	Date: *NOVEMBER 14, 1981*
Approved By: *KEN BROWN*	Date: *JANUARY 3, 1982*
Revisions:	Date:

Material Approval

Material Evaluation Approval No.	Date:
Manufacturer's Certification No.	Date:
Test/Research	Date:

Project Identification

Project No. *4004*	Date: *FEBRUARY 1, 1982*
Project Name *FIRST SAVINGS BANK*	Location: *BLUEBERRY, MICHIGAN*
Soils: *CLAY* Climate: *-25° TO 95°*	Rainfall: *30" PER YEAR*
Project Detail No. *12*	Sheet No: *5*

Field Evaluation Record

	Period	Not Acceptable	Poor	Good	Excellent	Acceptable
1	Construction				*✓*	*YES*
2	Five Year					
3	Ten Year					

Performance Problems

1	*NONE OBSERVED DURING CONSTRUCTION*
2	
3	

Reuse of Detail

Reuse Instructions:

1	Project No.: Project Name:	Project Detail No.: Date:	Sheet No.:
2	Project No.: Project Name:	Project Detail No.: Date:	Sheet No.:

Retrieval Descriptors

1. *PARAPET WALLS*	5.	
2. *ALUMINUM EXTRUSION FITS*	6.	
3. *MASONRY CONSTRUCTION*	7.	
4.	8.	

Figure 6-5 Data Card for Detail No. 000 316.

▷ *PARAPET WALLS*
  UF Dwarf Walls
  BT Guardrails
       Walls
  RT Crush Barriers
       Roof Parapets

▷ *ALUMINUM EXTRUSION FITS*
  BT Mechanical Assembly
  NT Contour Fits
       Dovetail Fits
       Slide Fits
       Snap Fits

▷ *MASONRY CONSTRUCTION*
  UF Masonry
  NT Blockwork
       Brickwork
       Glass Concrete
       Reinforced Masonry
       Stone Masonry

Figure 6-6 Descriptor Selection for Detail No. 000 316. These, descriptors are selected from the *Canadian Thesaurus of Construction Science and Technology*, July 1978.

Note.   Flexibility in selecting descriptors can be achieved by using other reference sources and by generating acceptable terms.

Example 2

▷ Descriptor Selection for Detail Number 000 163
▷ Stairs At Start—Figure 6-8 (page 126)
▷ Selecting Descriptors

Step 1. Study detail and identify critical areas of information being presented. See Section 6–1.3 for procedures.

Step 2. Select appropriate descriptor terms for indexing detail—Figure 6-7.

  a. Search existing thesauri for potential descriptors

  b. Generate new descriptors if required to define subject area of detail.

Step 3. Record descriptor terms for storage and retrieval on Detail Data Card and detail—Figure 6-9.—Page 127.

Results

▷ The descriptors that best describe the subject of detail are "Stairs" and "Stair Handrail."

▷ A Search of the *Canadian Thesaurus of Construction Science and Technology* and the *English Construction Industry Thesaurus* produce the following descriptors [11], [14]:

▷ | Stairs |

▷ | Stair Handrails |

▷ | Handrails |

Example 3

Multidisciplinary Use of Detail Banking System

Detail banking can be accomplished for all disciplines in the construction industry. Construction details generated by each of the following disciplines have potential for a detail banking system:

▷ architectural
▷ civil engineering
▷ mechanical engineering
▷ electrical engineering
▷ landscape architecture
▷ interior design
▷ industrial design.

Descriptor Selection
From Existing
Thesauri
▷ Canadian
▷ English

Functional parts of construction works
Parts for division, structure and circulation
Circulation elements
Vertical circulation elements
Stairs

J87290	▷ *STAIRS* = *Staircases*	
J87300	*Stair guards*	*Parts of stairs*
J87360	*Steps*	*ditto*
J87430	*Treads* = *Stair treads*	*Parts of steps*
J87500	*Risers*	*ditto*
J87570	*Tapered steps =*	*Types of steps*
	Winders	*by form*
J87620	*Flights*	*Parts of stairs*
J87690	*No turn stairs*	*Stairs by characterising parts*
J87760	*Quarter turn stairs*	*ditto*
J87830	*Half turn stairs*	*ditto*
J87900	*Three quarter turn stairs*	*ditto*
J87970	*Full turn stairs*	*ditto*
J88040	*Winding stairs =*	*ditto*
	Spiral stairs	
J88110	*Open well stairs*	*ditto*
J88150	*Non welled stairs*	*ditto*
J88170	*Continuous string stairs*	*ditto*
J88240	*Discontinuing string stairs*	*ditto*
J88310	*Flying staircases*	*Stairs by support*
J88380	*Straight stairs*	*Stairs by form in plan*
J88450	*Dog leg stairs =*	
	Single corkscrew stairs	
J88520	*Double corkscrew stairs*	
J88590	*Curved stairs*	*ditto*
J88600	*Pressurized stairs*	
J04710	▷ *Rails*	
J04730	*HAND RAILS*	*Rails by how used*

▷ *STAIRS*
  BT *Building Functional Elements*
  NT *Collapsable Ladder Stairs*
      Escalators
      Fire Escapes
      Folding Stairs
      Geometrical Stairs
      Half Turn Stairs
      Inner Stairs
      Ladders (Stairs)
      Open String Stairs
      Open Well Stairs
      Outside Stairs
      Overhanging Stairs
      Quarter Turn Stairs
      Return Stairs
      Sarasin Arched Stairs
      Service Stairs
      Spindle Stairs
      Spiral Stairs
      Straight Stairs
      Supported Stairs
      Walled Stairs
  RT *Access Ramps*
      Exterior Circulation Spaces
      Interior Circulation Spaces

▷ *STAIR HANDRAILS*
  BT *Guardrails*

    (a)   *(b)*

Figure 6-7 Descriptor Selection for Detail No. 000 163. The descriptors in (a) are taken from the *Canadian Thesaurus of Construction Science and Technology*, July 1978: Those in (b) are selected from the *English Construction Industry Thesaurus*, 1976.

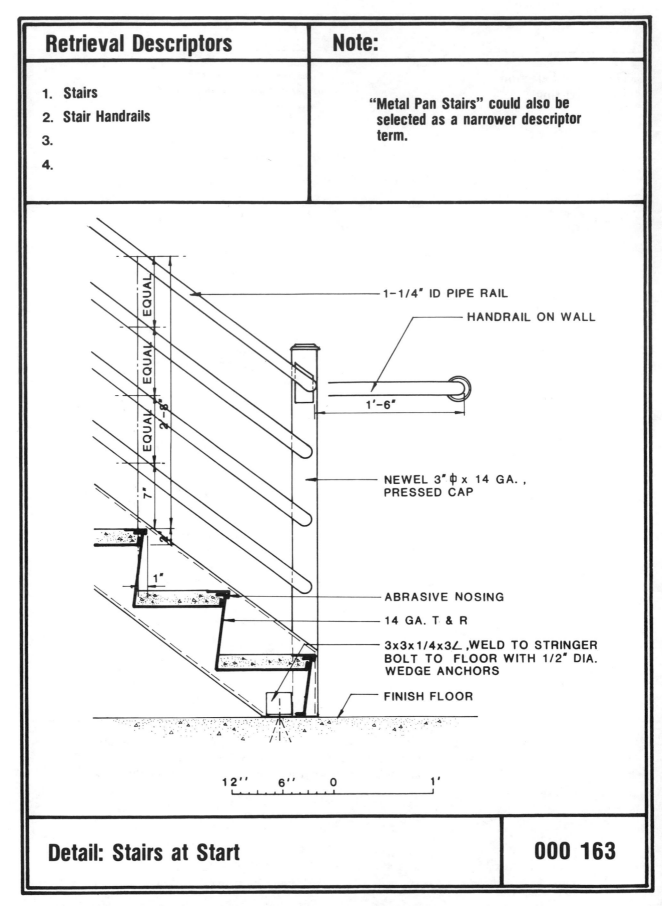

Figure 6-8 Retrieval Descriptors Identifying Stairs at Start Detail. (Detail Content: Gresham, Smith and Partners, Nashville, Tennessee.)

DETAIL DATA CARD

Bank Detail No.:	*000 163*
Detail Title/Description:	*STAIR & HANDRAIL*

Scale: *AS SHOWN*	Interior *✓*	Exterior

Development Data

Developed By: *PAUL SMITH*	Date: *JANUARY 12, 1979*
Approved By: *DAN WILSON*	Date: *MARCH 4, 1980*
Revisions:	Date:

Material Approval

Material Evaluation Approval No.	Date:
Manufacturer's Certification No. *501*	Date: *JUNE 1, 1980*
Test/Research	Date:

Project Identification

Project No. *0065*	Date: *AUGUST 1, 1980*
Project Name *BAYVIEW HOSPITAL*	Location: *CEDAR CREEK, IOWA*
Soils:	Climate:	Rainfall:
Project Detail No. *20*	Sheet No: *6*

Field Evaluation Record

	Period	Not Acceptable	Poor	Good	Excellent	Acceptable
1	Construction				*✓*	*YES*
2	Five Year					
3	Ten Year					

Performance Problems

1	*NO PRIME COAT ON STEEL*
2	
3	

Reuse of Detail

Reuse Instructions:

1	Project No.: Project Name:	Project Detail No.: Date:	Sheet No.:
2	Project No.: Project Name:	Project Detail No.: Date:	Sheet No.:

Retrieval Descriptors

1. *STAIRS*	5.
2. *STAIR HANDRAILS*	6.
3.	7.
4.	8.

Figure 6-9   Data Card for Detail No. 000 163.

An *electrical detail* has been selected to demonstrate

1. the use of detail banking in related construction industry disciplines,
2. the potential for detail reuse, and
3. the selection of appropriate descriptors.

▷ Descriptor Selection for Detail Number 000 005

▷ Control Wiring Diagram For Toilet Exhaust Fan—Figure 6-11 (page 130).

▷ Selecting Descriptors

Step 1. Study detail and identify critical areas of information being presented. See Section 6–1.3 for procedures.

Step 2. Select appropriate descriptor terms for indexing detail—Figure 6-10.

Descriptor Selection for Detail 000 005 from the Canadian Thesaurus of Construction Science and Technology *July 1978 Edition*

▷ *EXHAUST FANS*
    UF   *Heat Vents*
    BT   *Fans*
    PT   *Fume Vents*

▷ *TOILET FACILITIES*
    US   *Public Toilets*

▷ *WIRING*
    UF   *Electrical Wiring*
    WT   *Electrical Equipment*

Descriptor Selection for Detail 000 005 from the English Construction Industry Thesaurus *(1976 Edition)*

▷ *J*   *51460*   *EXHAUST FANS*
▷ *K*   *29710*   *TOILETS*
▷ *BS*   *4342*   *WIRING*

(Note to user: Identical descriptor terms can be found in two major construction thesauri. These sources provide the resource base for selecting descriptors to develop a national or in-house mini-thesaurus. The range of descriptors provides support for all disciplines of the construction industry.)

Figure 6-10. Descriptor Selection for Detail No. 000 005.

---

    a. Search existing thesauri for potential descriptors

    b. Generate new descriptors if required to define subject area of detail.

Step 3. Record descriptor terms for storage and retrieval on Detail Data Card and detail—Figure 6-12 (page 131).

Results

▷ The descriptors that best describe the subject of detail are "Exhaust Fans," "Toilet Facilities," and "Wiring."

▷ A search of the *Canadian Thesaurus of Construction Science and Technology* and the *English Construction Industry Thesaurus* produce the following descriptors [11], [14]:

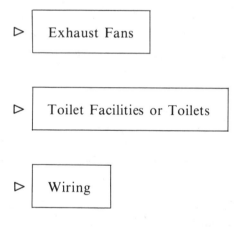

▷   Exhaust Fans

▷   Toilet Facilities or Toilets

▷   Wiring

6-2.1 Major advantages to indexing details with descriptor terms.

▷ Provides greater freedom in selecting appropriate descriptors to index details.

▷ Details can be easily cross-referenced through family relationships.

▷ Multidescriptor indexing provides many avenues to retrieve details.

▷ Increases the potential to access a specific detail.

▷ Allows greater flexibility to retrieve a group of related details.

6-2.2 Combining indexing systems.

Coordinating the CSI (Construction Specifications Institute's MASTERFORMAT) numbering system with the recommended "Descriptor Term System" can provide additional advantages to those firms who are linking details and specifications. By adding a CSI Division and Section number to related construction details, as described on page 95, it will further assist in coordinating specifications and drawings while providing another avenue for detail retrieval. Selected descriptor terms and specification section numbers will enable the designer and specifier to have ready access to both details and relevant specifications.

Automation systems have made it possible to link the production of graphic and text elements in the construction document preparation phase. By integrating information storage systems, it is possible to access reusable data instantaneously. A specifications reference number will provide an alternative identifier for selecting all details associated with a Division or Section. Organizations presently indexing details with CSI numbers can easily adopt the *"Descriptor Term System"* by simply selecting appropriate terms for each detail added to the banking system. Indexing details with both systems can provide the following benefits:

▷ Increases alternatives for storing and retrieving details

▷ Links details and specifications with descriptors and numbers

Example

Detail No. 000 163 on page 126 was indexed with the following descriptor terms:

▷  Stairs

▷ Stair Handrails

To index this detail under the CSI MASTER-FORMAT numbering system would require adding CSI No. 05510-0001 which identifies the Division and Section containing stair related information. By combining indexing systems, a search could be specified using the following alternatives:

1. Stairs
2. Stair Handrails
3. Stairs and Stair Handrails
4. Stairs, Stair Handrails and CSI No. 05510-0001
5. CSI No. 05510-0001
6. Stairs and CSI No. 05510-0001
7. Stair Handrails and CSI No. 05510-0001

6-3 SPECIFICATIONS STORAGE AND SIMULTANEOUS RETRIEVAL WITH DETAILS

Descriptor terms or key words provide the means for storing specifications information related to details selected for the banking system. Each descriptor selected for storing a detail can also be correlated with an appropriate section of specifications. This will encourage more effective coordination between drawings and specifications during the production process.

Detail retrieval can also recall the latest corresponding specification sections to be edited for the final contract documents. Efficiency in the total production of the contract documents can be achieved by integrating the storage of details and

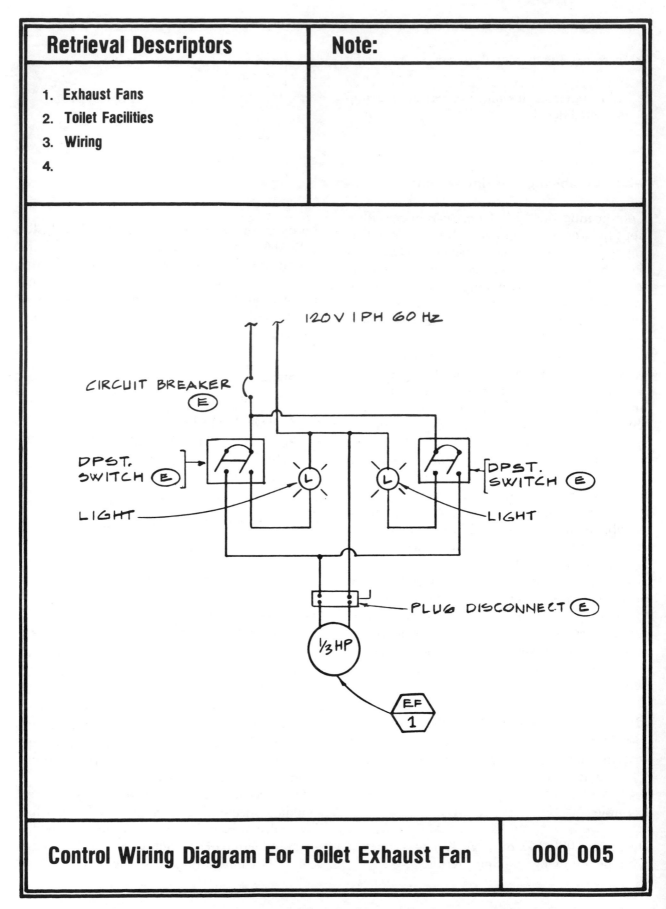

Retrieval Descriptors	Note:
1. Exhaust Fans 2. Toilet Facilities 3. Wiring 4.	

Control Wiring Diagram For Toilet Exhaust Fan | **000 005**

Figure 6-11 Retrieval Descriptors Identifying Control Wiring Diagram for Toilet Exhaust Fan. (Jacobs Architects, Pasadena, California)

DETAIL DATA CARD

Bank Detail No.:		000 005

Detail Title/Description: *CONTROL WIRING DIAGRAM FOR TOILET EXHAUST FAN*

Scale: *NONE*   Interior ✓   Exterior

Development Data

Developed By: *JIM ANDERSON*	Date:	*JUNE 20, 1980*
Approved By: *FRED JOHNSON*	Date:	*JULY 10, 1980*
Revisions:	Date:	

Material Approval

Material Evaluation Approval No.	Date:	
Manufacturer's Certification No. *105 - 6734*	Date:	*AUGUST 5, 1980*
Tests/Research	Date:	

Project Identification

Project No. *8001*	Date:	*SEPTEMBER 2, 1980*
Project Name *GREENVILLE ELEMENTARY SCHOOL*	Location:	*GREENVILLE, ILLINOIS*
Soils:	Climate:	Rainfall:
Project Detail No. *E-10*	Sheet No:	*6*

Field Evaluation Record

	Period	Not Acceptable	Poor	Good	Excellent	Acceptable
1	Construction				✓	*YES*
2	Five Year					
3	Ten Year					

Performance Problems

1	
2	
3	

Reuse of Detail

Reuse Instructions:

1	Project No.: Project Name:	Project Detail No.: Date:	Sheet No.:
2	Project No.: Project Name:	Project Detail No.: Date:	Sheet No.:

Retrieval Descriptors

1.	*EXHAUST FANS*	5.	
2.	*TOILET FACILITIES*	6.	
3.	*WIRING*	7.	
4.		8.	

Figure 6-12   Data Card for Detail No. 000 005.

specifications under a common descriptor term system. Standardization of terminology will aid the specifications writer and the draftsperson in coordinating specific items of work in the contract documents.

Example

In specifications storage, the selection of the stair detail shown in Section 6–2 will automatically retrieve the following specification sections:

Descriptor Term	Specifications
▷ Stairs	Division 5. Metals
	Section 05510. Metal Stairs

Part 1.   General

1.01 Work Included

A.   Furnish and install steel

▷ STAIRS

1.02 Related Work

A.   Section 03001. Concrete Work

Concrete fill in stair pans

B.   Section 05521.   Pipe and Tube Railings

Stair handrails at all open sides

1.03 Performance

A.   Construct stair assembly to support minimum live load of 100 lb per square foot

Part 2.   Products

2.01 Acceptable Manufacturers

A.   C & D Manufacturing Company

B.   Treads at Boiler Room and Transformer Vault stairways to be steel gratings manufactured by B & B Metal Products Company

Part 3.   Execution

3.01 Installation

A.   Install steel stairs and fittings in accordance with manufacturer's instructions

▷ Stair Handrails	Division 5. Metals
	Section 05521. Pipe and Tube Railings

Part 1.   General

1.01 Work Included

A.   Furnish and install steel
▷ pipe handrails including wall brackets at both sides of stairways

1.02 Related Work

A.   Section 05510.   Metal Stairs

Part 2.   Products

2.01 Materials

A.   Handrails to be constructed of $1\frac{1}{4}''$ nominal diameter standard steel pipe, all joints welded, ground clean and ends capped

B.   Brackets to be malleable iron #305

Part 3.   Execution

3.01 Installation

A.   Fasten securely to masonry or concrete walls.

6-4 METHODS OF STORING AND RETRIEVING DETAILS

Detail storage and retrieval systems should be designed around user needs and the level of information-handling technology. New advancements in automation have made it possible to minimize the manual activities associated with information processing. Before selecting a system, it is wise for the design professional to identify specific requirements and future expectations. The evaluation of system requirements and technological capabilities will help in structuring the most appropriate system for each design firm.

At the present time, one can choose from (1) manual, (2) semi-automated, and (3) automated systems for handling details (Figure 6-13). It is important for designers to evaluate each level before making a final selection. The procedures and recommendations for working with each system are outlined in this section.

1. Manual Systems

Details and data cards are indexed and filed manually in one or more of the following methods:

▷ Notebooks.
▷ Catalogs.
▷ Manuals.
▷ Files.
▷ Aperture cards, or microfilm systems.
▷ Card index,
  ○ Detail Data Cards,
  ○ Optical Coincidence Cards. (A system where numbers are compared by the relationship of holes punched in the file cards. Placing cards manually over a light source allows the user to rapidly retrieve multidescriptor information by matching accession numbers associated with selected descriptor terms.)*
  ○ Notched Card Systems.

*The most popular system in England [14].

File Detail Data Card under each descriptor used in indexing the detail. Each data card, as identified in Chapter 3, will receive a number that corresponds to the detail. The *data card* entered under selected descriptors serves as the locator source for each detail in the system. The search process is carried out by using the following steps:

1. Search thesaurus for appropriate descriptor terms.
2. Select the best descriptor terms for the final search.
3. Enter the *data card* file, and locate the selected descriptor terms.
4. Review *data cards* filed under selected descriptor terms for potential retrieval of details. Consider use of *optical coincidence card systems* for multidescriptor searches.
5. Select detail numbers that correspond to the details that are desired for retrieval.
6. Search detail catalogs, notebooks, or files for appropriated details.

Detail indexing can be done by using a single descriptor term or by selecting several terms that identify the detail content. The filing procedures are shown in Figures 6-14, 6-15 and 6-16 which demonstrate methods of entering the system.

2. Semiautomated System (Word Processors)

Details and data cards are processed for storage and retrieval with the aid of word-processing equipment. All information collected on Detail Data Cards can be stored in the word processor. The search process is conducted in the following manner:

1. Read subject area thesaurus, and select appropriate descriptor terms. Select one or more terms as required to define detail topic area.
2. Enter system by punching appropriate classification and accession numbers on keyboard:
   a. Entry is limited to 12 digits. Check capability of word processor before structuring the indexing system.

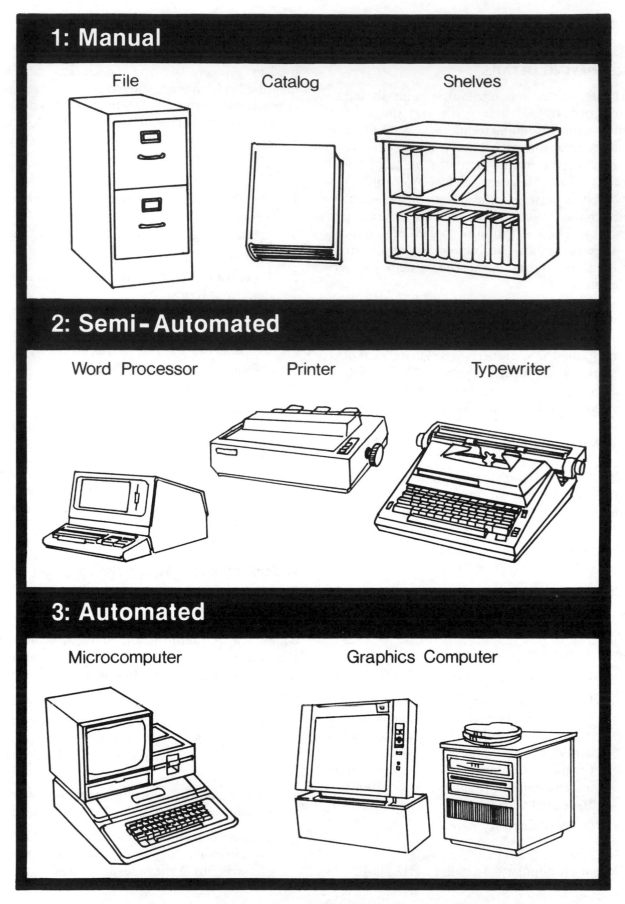

Figure 6-13 Methods of Storing and Retrieving Details and Data Cards.

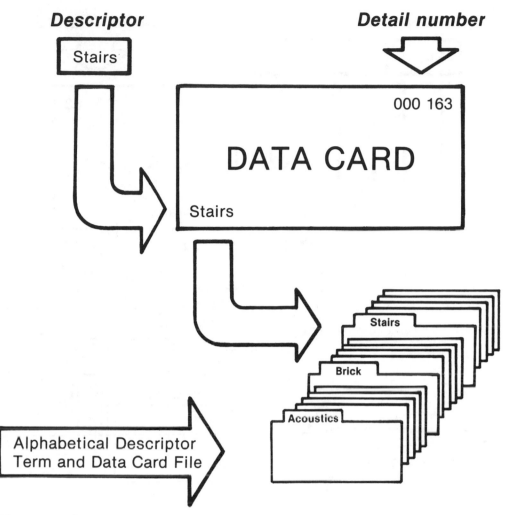

Procedure:

File Each Data Card Under the Appropriate Descriptor Term Selected for Detail

Figure 6-14 Single Descriptor Entries.

b. Storage and retrieval is by
    (i) numbers,
    (ii) numbers and letters, or
    (iii) descriptors and key words.

c. Multilevel retrieval uses up to seven levels at one time (seven descriptor terms can be entered for retrieval) and Boolean retrieval logic.

d. Storage on disks ranges from 70,000 to 600,000 characters per disk. Check system capacity.

3. Retrieve Detail Data Cards on screen and review each card:

    a. Retrieve all cards associated with selected descriptors.

    b. Obtain printout of list if required for further search.

4. Record classification and accession numbers associated with details selected for retrieval. Select desired details after reviewing information on Detail Data Cards.

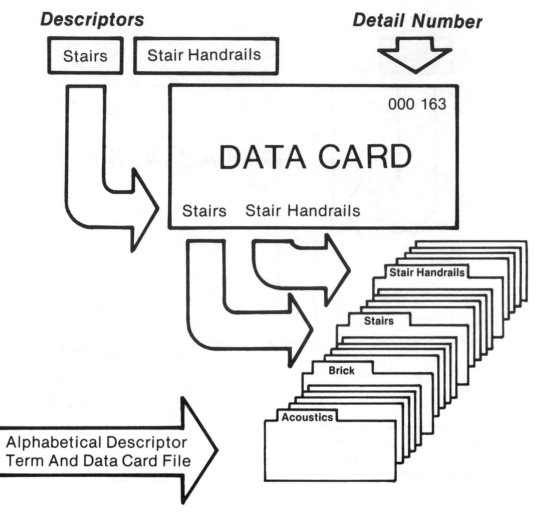

Procedure:

File a Copy of Data Card Under Each Descriptor Term Selected for Detail

Figure 6-15   Multidescriptor Entries.

5. Retrieve details from manual storage systems by searching descriptor categories and numbering system.

3. Automated System (Computers and Microcomputers)

Details and data cards are processed for storage and retrieval from a computer data bank. All written and graphic information related to details can be stored in the computer.

File Detail Data Cards and detail under each descriptor used for indexing. In search ask for a printout of only those details that have the selected descriptor. Computer searches retrieval terms selected and prints out all numbers under selected term. When multidescriptors are selected, the computer prints out only those numbers that are common for each term selected.

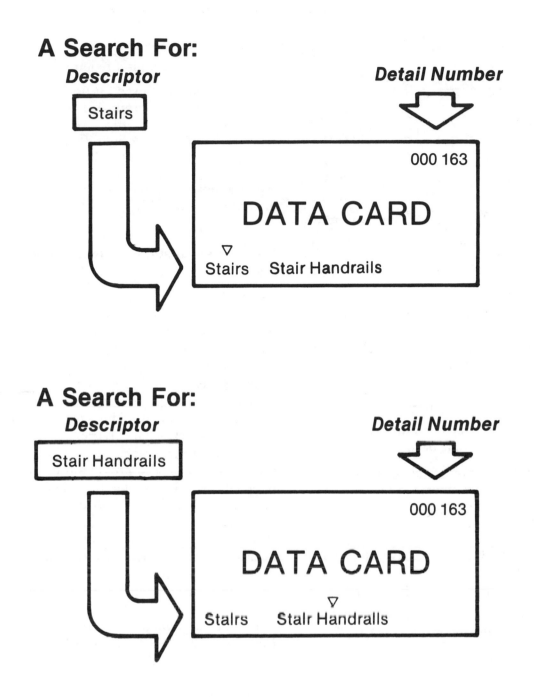

A Search For:

Descriptor

Stairs

Detail Number

000 163

DATA CARD

▽
Stairs   Stair Handrails

A Search For:

Descriptor

Stair Handrails

Detail Number

000 163

DATA CARD

▽
Stairs   Stair Handrails

Procedure: A search under each descriptor identifies detail number 000 163

Figure 6-16 Multidescriptor Search.

Example

Descriptor Selection	Detail Numbers
Walls	1, 2, 3, 4, 6, 10, ⑯
Acoustical Walls	12, 14, ⑯, 25, 30
Audio Video Laboratories	5, 8, ⑯, 20, 35

Detail number ⑯ uses all retrieval terms, therefore, the computer printout will identify only number ⑯. The Boolean retrieval process provides the searcher with a selective detail search. The search process involves the following steps:

1. Read subject area thesaurus and select appropriate descriptor terms:

    a. Select one or more terms, as required, to define detail topic area.

    b. Determine which descriptors should enter system (single or multi).

2. Enter system by punching appropriate descriptor terms on keyboard.

3. Retrieve Detail Data cards on screen and review each card:

    a. Retrieve all cards associated with selected descriptor.

    b. Obtain printout of list if required.

    c. Retrieve details directly on screen if data card information is not required.

4. Record accession numbers associated with details selected for retrieval:

    a. List numbers for each descriptor printout.

    b. Select those required by reviewing information on Detail Data Card.

5. Have a computer recall selected details by punching in accession numbers recorded:

    a. Project detail on screen.

    b. Review for required information.

    c. Revise if necessary.

    d. Obtain printout if required.

6. Have detail retrieved for branch offices that have computer terminal. Details can be retrieved by branch offices, contractor offices, or field offices, depending on terminal locations.

Applications of all three systems are shown in the matrix in Figure 6-17.

6-4.1 User variations in steps to detail banking.

Detail Data Card development can be removed from the input and retrieval steps, if historic data is found unnecessary. *However, a data card is recommended to gain full advantage of an evaluation system.* It provides a record of detail improvements as well as directions for proper usage.

Selecting descriptors is critical to the storage and retrieval process. This step provides the means for indexing and retrieving a detail by appropriate descriptors as opposed to one fixed title. This system also provides users maximum flexibility to choose the best descriptors for indexing and retrieving each detail selected for banking.

The basic steps for storing and retrieving details have been identified to assist users in developing effective systems. Variations in these procedures can be introduced, when required, for specific user needs and goals. However, the major steps for detail storage and retrieval are shown in Figure 6-18 and 6-19 (pages 140 and 141).

6-4.2 Processing details for banking.

▷ Develop original and one copy of detail.

▷ Check if detail has already been processed.

▷ Make initial screening and evaluation of detail.

▷ Select detail appropriate for banking system.

▷ Does detail merit inclusion.

▷ Assign accession number to detail.

▷ Select descriptor terms.

Methods of Storing and Retrieving Details

	MANUAL	SEMI—AUTOMATED	AUTOMATED
Numbering	Level 1 or 2	Level 1 or 2	Level 2
Equipment	Typewriter	Word Processor	Computer System
Retrieval Method	Single or multidescriptor Note: single is easier	Single or Multi— Descriptor	Single or Multi— Descriptor
Detail Accessibility	Original or Copy of Detail	Original or Copy of Detail	Graphic projection or copy of detail
Revisions	Redraw and Photograph	Redraw and Photograph	Revise on CRT or redraw and modify a printout
Selection Of Detail	Select descriptor from Thesaurus	Select descriptors or descriptor from Thesaurus	Select descriptors or descriptor from Thesaurus
Entry to Data Cards	Enter Detail Data Card file and review entries under selected descriptor terms	Punch in descriptor term numbers and review all Data Card entries under a given descriptor	Punch in descriptor term and review all Data Cards and under selected descriptor terms
Entry to Detail Bank	Record desired detail numbers from Data Cards and search detail file	Record desired detail numbers from Data Cards and search detail file	Record desired detail numbers from Data Cards and punch in numbers for graphic projection on screen.

* A Level One or Two numbering system can be used with any method of storage and retrieval.
However, to achieve the greatest effectiveness, a Level Two Numbering System is recommended
for all methods of Detail Banking. See page 144.

Figure 6-17 Application of Storage and Retrieval Systems—a matrix overview.

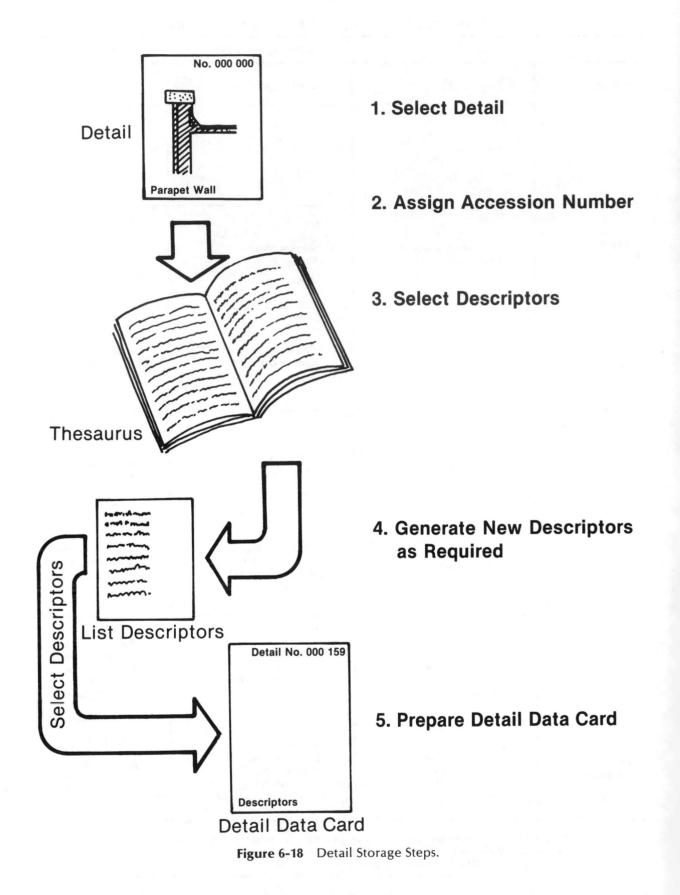

Figure 6-18   Detail Storage Steps.

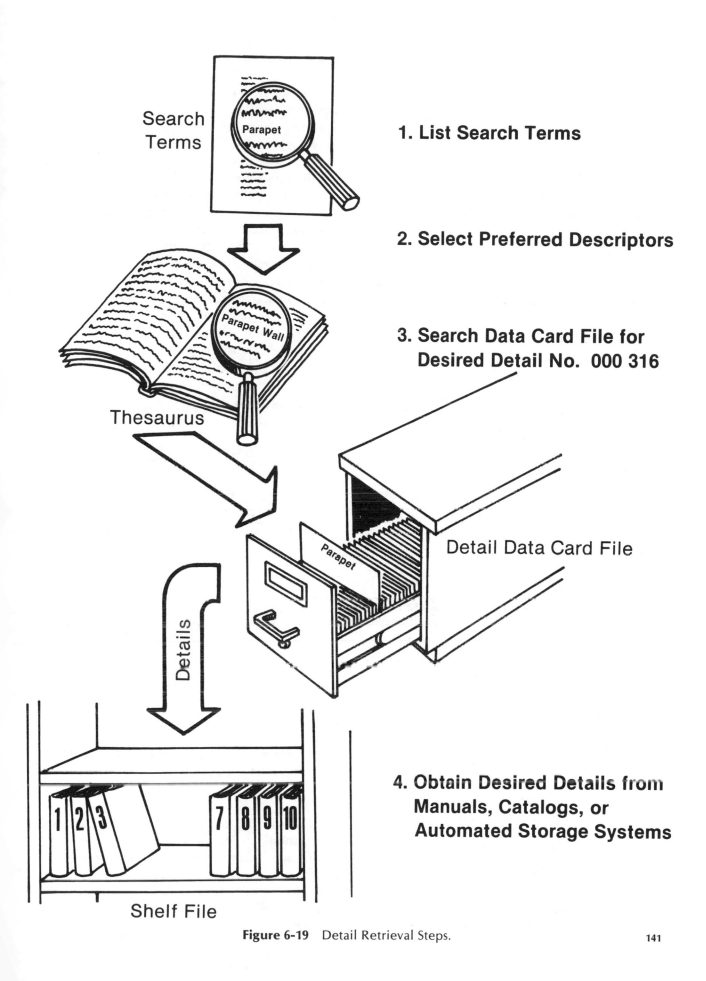

Search Terms

1. List Search Terms

2. Select Preferred Descriptors

Parapet

3. Search Data Card File for Desired Detail No. 000 316

Thesaurus

Parapet Wall

Detail Data Card File

Parapet

Details

4. Obtain Desired Details from Manuals, Catalogs, or Automated Storage Systems

Shelf File

Figure 6-19 Detail Retrieval Steps.

141

▷ Conduct quality review of detail and descriptor terms.

▷ Prepare Detail Data Card.

▷ Submit detail and Detail Data Card to librarian.

▷ Prepare for storage and retrieval.

▷ File detail and Detail Data Card.

▷ Monitor detail performance and obtain field feedback.

▷ Revise and update detail.

6-5   HOW TO DEVELOP NUMBERING SYSTEMS

Effective numbering systems play an important role in the storage and retrieval of information. Letters and numbers can be selected to represent categories and classifications for each item filed in a series of related materials. Numbering systems can be developed at two levels for storage and retrieval:

▷ *Level One Number* = RS-1-0253. The letter and numbers serve as a symbol in the first level of search to an information system. Selecting the right numbers is critical in obtaining the appropriate information:

▷ *Level Two Numbering* = Descriptor Term = Chimney. Detail number = 0.140 *THE RECOMMENDED NUMBERING SYSTEM.* The descriptor term serves as the *first level* entry into the information system. A detail accession number provides a *second level* of search to locate a specific item of information.

To adapt the U.S. CSI/UCI and English CIT information programs for detail banking systems, it is necessary to create an expanded number for each item stored in a specific division or facet. The numbering system for both programs serves as the first level of entry into the storage system. An example of the numbering follows:

Level One Numbering System

System	Existing Number	Requirements for Detail Banking
CSI/UCI	05300	05300–001
CIT	J96010	J96010–001

The descriptor term system for information handling is usually associated with a second-level numbering system. Each detail entered into this system is assigned a drawing number or accession number. The numbering system is open ended, and drawings can be added to this system indefinitely. Thus the first order of retrieval in this system is the subject area descriptor. The second level of retrieval is associated with the detail number given to the drawing when it is entered into the system. Detail numbers are consecutive and have no association with subject categories, thus allowing the total system to be flexible to expand as needed. This type of numbering system can best be used with the *Canadian* or *ERIC Thesaurus*, or with an individually created thesaurus of construction terminology. The following example demonstrates the use of an accession numbering system:

Level Two Numbering System

System	Existing Number	Requirements for Detail Banking
Canadian	None	Descriptor and Accession Numbering 000 100
ERIC	None	Descriptor and Accession Numbering 000 050

6-5.1   How to determine which numbering system is best for you.

User needs and information-handling requirements should generate the determinants for establishing numbering systems, classifications, categories, and other special requirements for both existing and new data-filing programs. The following analysis of two numbering systems identifies their key elements.

Level One System

CSI/UCI. Based on 16 Divisions of the specifications:

Number = 10   700

Division Number | Item Number

CIT. Based on classification of subjects listed under each facet:

Number = J   96010

Classification of Facet | Item Number

In a *level one numbering system*, the critical elements are composed of the following:

Number =	RS	1	0253
Break-down analysis	= General classification of subject area	Specific item number for each topic	Detail numbering

Level Two System

Level two numbering is less complex and is not dependent on categories or classifications. The descriptor term serves as the first level of entry into the data bank. Descriptor terms are associated with detail accession number in order to identify every detail where a descriptor was used. Accession numbers are identified for each detail as they are accepted into the detail bank. They serve as detail locator numbers for descriptor reference and detail selection. The following descriptor demonstrates how a detail number is used with a specific term:

▷ Descriptor term = **Architectural Barriers.**

▷ Detail numbers where descriptor is used = 0001, 0009, and 0216.

Each number is used to identify the detail as it is added to the system. It can be considered as an accession number since all details are numbered sequentially as information is added to the data bank. *This numbering system is simple, direct, and flexible to add data indefinitely without changing the format.* Users of the system must become concerned with subject descriptors rather than symbolic numbering systems for retrieval. The number of digits used to make up the total detail number is dependent on:

1. A decision to predetermine the maximum number of digits to allow the system growth. For example, a six digit capacity

   000   000
   000   001
   000   002

is recommended to control number spacing.

2. A decision to start with number one and add detail numbers indefinitely in a consecutive order, for example,

   1,   2,   3,   4,   5,   etc.

Letters can also be added to the numbering system to identify an organizational code, or as an abbreviation for the company name. This type of identification is necessary when different forms of information-handling programs are being utilized in the same firm. The letters selected will remain constant thoughout the system. For example, an architectural engineering firm by the name of Architects and Engineers Collaborative might decide to show ownership of the information system by creating the following alphanumeric number format:

AEC   000   000

Firm Abbreviation   6 digits allowing for 999,999 details

For the purpose of achieving a constant number format, it is desirable to select a fixed number of letters and digits at the origin of a program. A predetermined digit number and spacing allows for greater consistency in both manual and automated systems.

Numbering System Components

Level 1.   Classifications and categories provide a format for details to be grouped and numbered sequentially by accession number:

1. Classification or category of descriptors = alpha.
2. Descriptor term/item number = number.
3. Detail acquisition = number.

Example

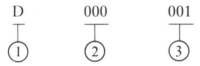

Level 2.   Descriptor terms and consecutive accession numbers assigned to details in order of entry to bank:

1. Classification = alpha for overall system-firm abbreviation.
2. Detail acquisition = number sequentially for overall system.

Example

6-5.2   How to select a numbering system.

A thorough analysis of the organization's information-handling needs and objectives will help direct one's selection of the final numbering system. By asking the following questions, a design professional can choose the best system:

1. Is it desirable to select an existing information-handling format like the CSI/UCI, the English CIT, or the Canadian construction terminology system?
2. Is it desirable to obtain documents from existing information-handling programs and use them as a reference base for creating a new system?
3. Is it desirable to limit a new system to a small-scale and manual operation?
4. Is it desirable to design a new system for flexibility and automation?
5. Is it desirable to file all details under specific categories in manuals or a notebook?

Level One Numbering

For classifications and categories of details select a level one system, such as shown in Figure 6-20:

▷ When choosing to work with existing information-handling programs, like the CSI/UCI, or the English CIT.

▷ When choosing to store details manually in specific categories located in notebooks, catalogs, or files.

Reasons

▷ The existing categories and numbering systems will allow one to group related details according to the established standards.

▷ Freedom in classification and category design allows one to structure the system according to individual needs.

Level Two Numbering

For descriptor terms and accession numbers select a level two system, such as shown in Figure 6-21:

▷ If choosing to use a construction language (descriptor terms) as the primary level of indexing, searching, and retrieving information.

▷ If choosing to store information in automated systems like the word processor and computer.

▷ If choosing to have unlimited flexibility to store information and expand system.

Reasons

▷ Construction language can be expanded to accept a variety of information areas.

▷ Retrieval is dependent on selection of appropriate descriptor terms to obtain a printout of stored details.

▷ Search process is language dependent and therefore eliminates a need for classifications and categories.

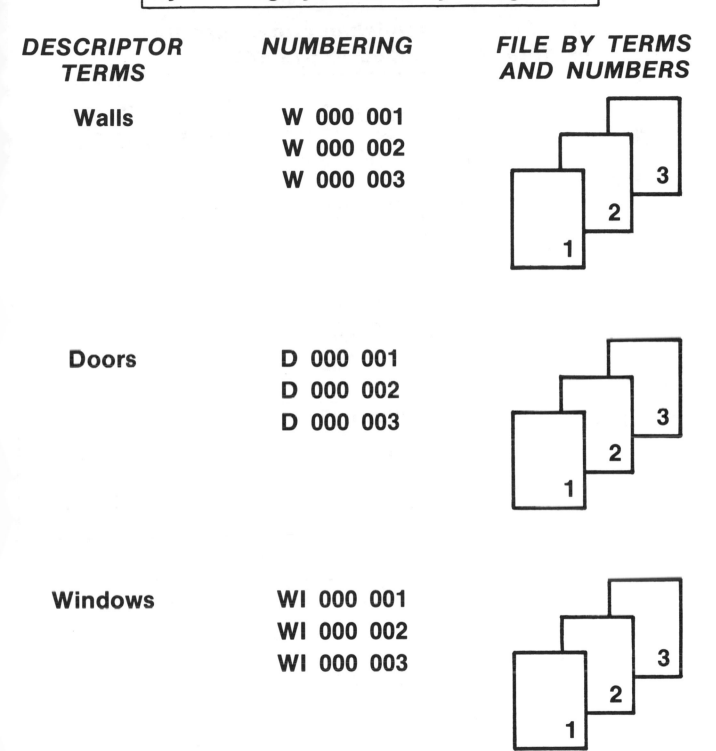

DESCRIPTOR TERMS	NUMBERING	FILE BY TERMS AND NUMBERS
Walls	W 000 001 W 000 002 W 000 003	
Doors	D 000 001 D 000 002 D 000 003	
Windows	WI 000 001 WI 000 002 WI 000 003	

Figure 6-20 Level One Numbering: Filing Detail by Terms and Alphanumeric Numbers.

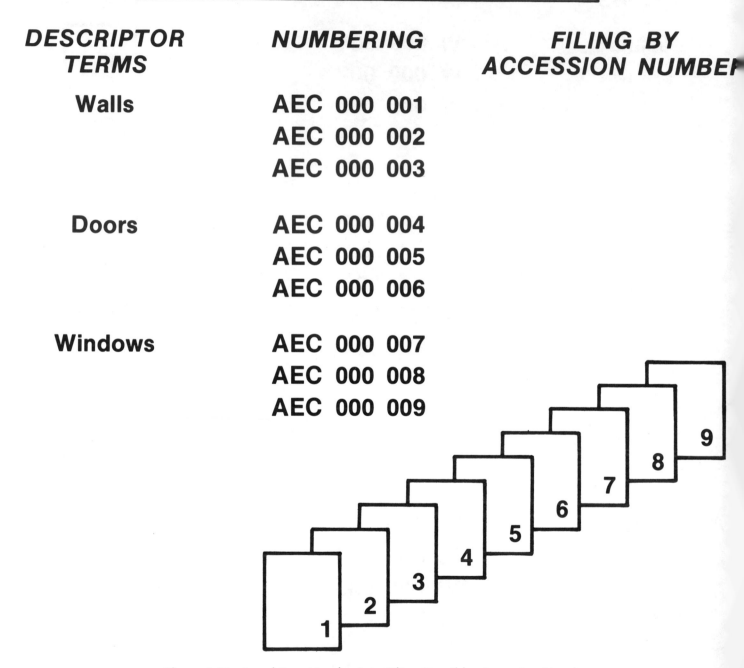

Type A

Numbering established consecutively as details are added to system.

DESCRIPTOR TERMS	NUMBERING	FILING BY ACCESSION NUMBER
Walls	AEC 000 001	
	AEC 000 002	
	AEC 000 003	
Doors	AEC 000 004	
	AEC 000 005	
	AEC 000 006	
Windows	AEC 000 007	
	AEC 000 008	
	AEC 000 009	

Figure 6-21   Level Two Numbering: Filing Detail by Accession Numbers.

▷ Cross-referencing is accomplished by selecting a multidescriptor search and retrieval process.

Optional Filing Systems for Level Two Numbering

If you are working manually and predict the use of automation in the very near future, consider using only descriptor terms and identify all details entered by a consecutive accession numbering system. This concept does not require you to predetermine categories of descriptor terms. You can choose from two different filing methods (Figures 6-21 and 6-22):

Type A.   File all details consecutively by accession number:

▷ **Example**

1,   2,   3,   4,   5,   etc.

Type B.   By descriptor grouping with numerical gaps:

▷ **Example**

1,   6,   8,   9,   15

One may want to create a subnumbering system for each descriptor grouping until automation program is operational. A temporary type B filing of details will provide the overall benefits of a level two numbering system plus the advantage of grouping details for manual use. Figure 6-23 (page 149) demonstrates numbering system applications for detail banking.

6-5.3   Advantages and disadvantages of numbering systems.

Level One Numbering

Advantages:

▷ Provides for grouping similar materials into one notebook, shelf, or other type of compartment for a single category of materials.

▷ Allows searcher to browse or scan a series of similar materials.

▷ Makes it easy to tell how many details are stored under one descriptor term.

Disadvantages

▷ Does not provide for easy retrieval of materials using a multidescriptor search.

▷ Is basically limited to a single descriptor search process.

▷ Must number or associate with descriptor to detect what category for storage.

▷ Requires a numbering system for each detail category.

▷ Searcher may become confused with classifications and categories of numbering system.

Level Two Numbering

Advantages

▷ Provides for *multidescriptor indexing* and retrieval.

▷ All documents are assigned an accession number upon receipt and entry.

▷ All numbers in the system are in a consecutive series.

▷ Allows a simple number system for all entries.

Disadvantages

▷ Does not allow for easy grouping of similar materials in manual storage.

▷ Does not provide for easy scanning or browsing of related materials.

▷ It is more difficult to determine the number of details in one category when used manually.

6-6   EVALUATING METHODS OF INDEXING AND STORING DETAILS

Indexing serves as a guide to facilitate the reference of information. It is important to develop it

Optional Filing System — Type B

Filing system provides for a manual cataloging of details by descriptor term prior to establishing storage in automation system. Converting system to automation would simply merge all details into the consecutively organized format shown in Type A.

DESCRIPTOR TERMS	NUMBERING	FILE BY TERMS AND ACCESSION NUMBER
Walls	AEC 000 001	
	AEC 000 004	
	AEC 000 008	
Doors	AEC 000 002	
	AEC 000 005	
	AEC 000 007	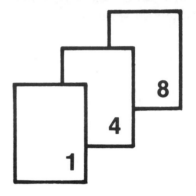
Windows	AEC 000 003	
	AEC 000 006	
	AEC 000 009	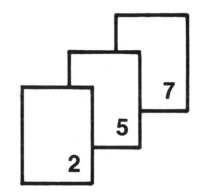

Figure 6-22 Level Two Numbering: An Optional Filing System by Terms and Accession Numbers.

Numbering System Applications

Level One — a system that groups details into classifications and categories
established by descriptor terms and numbers.

1	2	3
Thesaurus Terms receive an alpha code and number as they are added to the system	Detail Data card receives alpha code from descriptor plus the number of term and a number for the detail being added to the bank	Detail receives the same alpha code and number as on Detail Data Card

Level Two — a system that relates descriptor terms to accession numbers on
details filed in a consecutive series.

1	2	3
Thesaurus Terms serve as identifiers on each detail	Detail Data Card receives number of detail generated Example: 1, 2, 3, etc.	Details numbered as they are added consecutively to Data Bank Example: 1, 2, 3, etc.

Figure 6-23 Numbering System Applications.

using one approach—numbers, symbols, terms, or codes—in as simple a manner as possible. A researcher's time is greatly dependent on the method selected to index information. The preceding descriptions of descriptor terms and numbering systems provide the basis for creating an effective detail banking system.

It is important to distinguish the two methods of indexing and storing details. The first method was described in the *level one numbering system*. In this method details are grouped by alphanumeric categories related to subject areas. Each detail is filed as part of a family or grouping in notebooks, files, or special manuals created for different categories of details. For example, all "interior

partition" details would be indexed and stored under a common broad classification of "partitions."

This approach to indexing requires the indexer to predetermine detail groupings before the search process takes place. The search process requires careful identification of the desired categories of related details before entering the system. *As categories of information grow, this system can become more complex and difficult to manage.* The groupings of related details prior to a search can result in indexing the same subject areas under several different categories. For example, acoustic insulation details could be indexed under the following major categories:

▷ Residential

▷ Hospitals

▷ Factories

▷ Auditoriums

The second method of indexing was identified by the *level two numbering system*. In this method details are indexed by descriptor terms and given accession numbers as they are added to the system. It is important that every detail receive an accession number when it is selected for the banking system. Details are usually filed in consecutive order in files, notebooks, manuals, or in automated systems. No direct relationship is established between details during the indexing and storage process. For example, a Detail Data Card with the descriptor term "windows" is given an accession number 000 155. This would indicate that the detail number 000 155 contains information on windows. However, a detail number 000 156 could contain completely different information depending on the selection process. Figure 6-24 shows this relationship and

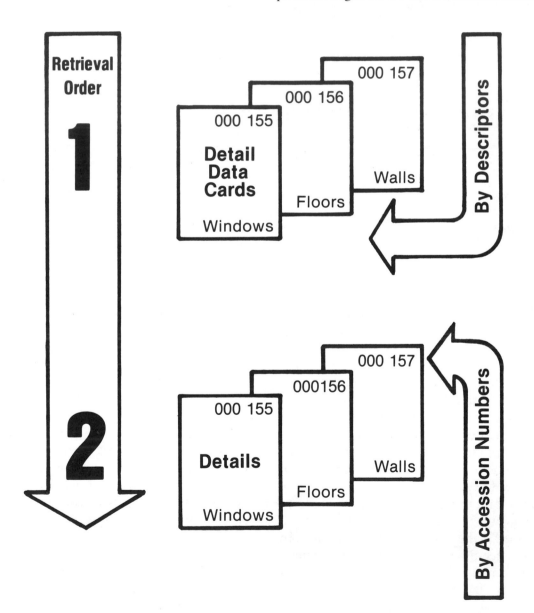

Figure 6-24   Retrieval Order for Data Cards and Details.

the order of data retrieval for a manual system. The search process requires selection of descriptor terms that outline the overall content of the information requested. Several descriptor terms can be identified and brought together to formulate the search process.

This method allows the researcher complete flexibility in formulating a search program to satisfy specific goals and objectives. The search is not restricted to categories or classifications. The key advantage to this method is that the searcher is free to specify the subject areas to be utilized in the search process. No prior coordination of categories or terms is required before formulating the search. Each descriptor term is treated on an equal level during the search process. Therefore no subordinate categories of descriptors can create alternative choices for filing details. The overall advantages of the second method of indexing are:

▷ Details are not restricted to categories.

▷ Flexibility in structuring the subject area of search.

▷ Ease of understanding overall system.

▷ More accurate indexing and retrieval.

▷ Increased chances of finding relevant details.

▷ Process is structured for automation.

6-7 EFFECTIVE METHODS OF SEARCHING AND RETRIEVING DETAILS

The detail search process is directly linked to the construction language generated for the detail banking system. As described earlier, details are indexed with descriptor terms and accession numbers for effective storage and retrieval. Each descriptor term serves as an identifier of basic concepts presented in the detail. The search process is built on the same basis as the indexer's guidelines for deciding which concepts (descriptor terms) best represent the subject of a particular detail.

A *level two system* is structured so that the descriptor term is the first level of entry into the detail bank. To initiate a search requires identification of the key concepts associated with the researcher's design problem. Once the concepts are identified, they should be documented and used to search the construction thesaurus for the appropriate descriptor terms. For example, if a design calls for a detail to solve a window problem, the searcher might select the following terms for the search process:*

WINDOWS

UF	Fenestration
RT	Building Design
	Climate Control
	Daylight
	Glare
	Glass Walls
	Illumination Levels
	Lighting
	Ventilation

GLASS WALLS

SN	Walls Consisting Largely of Windows
UF	Window Walls
RT	Building Design
	Classroom Design
	School Design
	Windows

VENTILATION

RT	Air Conditioning
	Air Conditioning Equipment
	Air Flow
	Chimneys
	Climate Control
	Controlled Environment
	Design Needs
	Exhausting
	Fuel Consumption
	Heating
	Lighting
	Mechanical Equipment
	Physical Environment
	Temperature
	Thermal Environment
	Windowless Rooms
	Windows

□

The actual search can be conducted as either a **narrow** or a **broad** search. A narrow search for the example problem would be to retrieve only the information where "Windows," "Glass Walls,"

---

*Terms selected from the *Thesaurus of ERIC Descriptors* [13].

and "Ventilation" are used as descriptor terms. If too little information is retrieved, the search must be broadened.

There are two methods of broadening a detail search. The first method would be to eliminate some of the terms that must concur on the same detail. By removing "Glass Walls" and "Ventilation"' from the search, it would broaden the retrieval process to include everything filed under the term "Windows." Through the process of subtraction or addition of individually used descriptor terms, the searcher can broaden a search.

The second method of broadening a detail search expands the scope of each descriptor term and utilizes related terms. Using the thesaurus to go from narrow terms to broad terms enables this process to develop for each family of descriptor terms. For example, to broaden the search for "movable partitions" one would also include a search of the (BT) "Space Dividers":

Descriptor Terms.   MOVABLE PARTITIONS

Sn	Interior Walls
	Can be readily
	moved.
UF	Folding Partitions
BT	Space Dividers
RT	Flexible Classrooms
	Flexible Facilities
	Prefabrication
	Space Utilization

Broad and narrow terms are generically related and therefore can lead searchers to other valuable search areas.

To aid the searcher in selecting a broad or narrow search, it is desirable to maintain an updated count of the details using a specific descriptor. For example, the descriptor terms would be displayed in a printout as follows:

Number of Details Using Term	Descriptor Term
50	Space Dividers
15	Movable Partitions

Identifying the number of entries helps the searcher formulate the search menu more accurately. The number of entries cited provides the searcher with clues as to how broad or narrow to make the search.

The process of organizing a search is greatly dependent on the searcher's ability to understand the construction language and to select appropriate retrieval terms. Many options are open to combine descriptors for structuring a search. Several search considerations impact on the effectiveness of information retrieval:

▷ User's ability to express requirements clearly.

▷ User's interpretations of system capabilities and limitations.

▷ Capacity of the system's vocabulary to express the user's needs.

▷ Ability of the searcher to recognize and cover all possible approaches to retrieval.

▷ Level of search strategy adopted.

6-7.1 Cross-referencing details.

The family relationship of terminology associated with each descriptor is a valuable aid to cross-referencing information. A genus and species relationship establishes a family structure for developing the framework for each descriptor. For example, an analysis of the term "Dining Facilities" indicates that the Broad Term referenced is "Facilities." Following the Broad and Narrow Term is a series of Related Terms that are closely connected in concept and remind the searcher of existing relationships among descriptors. The extent of cross-referencing within a generic group can be established by controlling the development of each descriptor. An effective cross-referencing system can be achieved *first* by developing categories and families in the selection of descriptor terms and the construction of a thesaurus.

EXAMPLES

THERMAL ENVIRONMENT

SN   Related to Combined Effects of Radiant Temperature, Air Temperature, Humidity, and Air Velocity

BT   Physical Environment

RT   Air Conditioning
Air Condition Equipment
Building Design
Climate Control
Climatic Factors
Controlled Environment
Environmental Influences
Fuel Consumption
Heating
Human Engineering
Humidity
Interior Design
Pollution
School Environment
Solar Radiation
Standards
Temperature
Ventilation

UTILITIES

UF   Electric Utilities
Gas Utilities
Public Utilities
Water Utilities

BT   Services

RT   Communication
Electrical Systems
Fuels
Heating
Kinetics
Lighting
Sanitary Facilities
Sanitation
Telephone Communication Systems

Second by using multidescriptor term indexing and retrieval methods, the searcher selects related descriptors to cover the scope of the problem area effectively. A Boolean-type search strategy will enable the searcher to cross-reference terms to achieve a narrow or a broad search. For example, the relationship between the following terms will allow the searcher to become very selective in the search for "interior partitions":

▷ Interior Partitions

▷ Acoustical Partitions

▷ Libraries

By cross-referencing all three terms, the searcher can obtain only the "Acoustical Partition" type details rather than all "Interior Partition" details.

6-7.2 Boolean retrieval system.

The Boolean retrieval system is a process whereby information that is indexed with subject area descriptors can be retrieved by selecting descriptors associated with specific items of information desired. A Boolean retrieval process works simultaneously with several descriptors to retrieve only those documents that have been identified with the indicated descriptor terms. This system allows the user great flexibility in selecting information while leaving the complicated search process to the communication system. Retrieval by selecting key words and subject areas through the use of a Boolean retrieval selection system has provided users with an effective means of accessing information with the least problems.

Boolean retrieval programs can be written to accept search requests in the form of descriptor listings and logical term operators such as combinations of "and," "or," and "not." For each descriptor in the retrieval request, a corresponding set of detail numbers are processed as a set under Boolean algebraic rules. The output is a printout of all details whose descriptor sets satisfy the retrieval request.

This type of retrieval system provides the searcher with an opportunity to prescribe the kind of details desired, selecting from a variety of descriptors only those terms closely related to the subject area of the problem. The following request will demonstrate the process associated with Boolean retrieval.

Problem Statement

A school requires a detail of a movable type partition that has a sound absorbing surface.

Procedure

The searcher generates the following list of potential search terms by carefully studying the descriptors identified in a construction thesaurus:

Subject Area	Descriptors
Construction Element, Product or Material	Partitions Interior Partitions Acoustical Partitions Movable Partitions
Facility Type	Churches Hospitals Schools Factories

Selecting Descriptor Search Terms

To reduce the number of details identified, the searcher decided to prescribe a "narrow search" by selecting all the following terms:

▷ Acoustical Partitions
　　　　and
▷ Movable Partitions
　　　　and
▷ Schools

If a broader search is desired, one could prescribe the following descriptor combination:

▷ Interior Partitions or Acoustical Partitions
　　　　and
▷ Movable Partitions
　　　　and
▷ Schools

Complete freedom is allowed the searcher in defining the limits of an individual search. The Boolean retrieval system provides for extreme flexibility in creating a variety of search term combinations that can aid the searcher in finding specific information in the shortest possible time. Word processors and computers can be programmed to store and retrieve information, using Boolean retrieval logic.

6-8　WHY CHOOSE THE RECOMMENDED MASTER SYSTEM

Feedback from many qualified sources provide positive implications for structuring a construction language and using it as a framework for developing information-handling systems. The scope of a standardized language can generate the basis for a "universal" or "master" system. Greater benefits will be derived from a language-based system as the design profession expands the use of automation. The following implications highlight the reasons for developing a master system:

▷ Effective information systems in other disciplines use a *controlled vocabulary and Boolean retrieval logic.*

▷ A 1978 survey of the characteristics and performance for 14 U.S. information systems in different disciplines shows

　　1.　11 out of 14 use a subject related vocabulary

　　2.　7 out of 14 indicate a need to further standardize a vocabulary [8].

▷ Studies show that a "descriptor system" provides better access to stored information.

▷ A language-based system will provide total flexibility to keep pace with a changing construction technology.

▷ A construction vocabulary and thesaurus system can be easily understood by its users.

▷ The user works with a common and standard language of the construction discipline.

▷ The system provides multidescriptor indexing for more selective information retrieval.

▷ The searcher enters system with descriptors that have a direct relationship to the subject area under search.

▷ Descriptor terms are added to the thesaurus as required by user needs and changes in construction technology.

▷ The family relationships of descriptors provide for effective cross-referencing of information.

▷ The retrieval system requires minimum involvement on the part of the searcher.

▷ The automated retrieval eliminates direct interaction of the user with the complexity of the search process.

▷ A descriptor and Boolean retrieval information system develops many alternatives for indexing information. Details can be indexed, stored, and retrieved by

1. building type, by
2. material composition (identification of more than one material when required),
3. object or element,
4. other descriptor subject areas as defined by user needs.

▷ The recommended master system has many advantages over other systems as stated in Section 5-3.4; however, other systems can also be used in conjunction to

1. maintain continuity during new system development,
2. supplement the overall master system for special uses,
3. achieve greater flexibility in managing information flow.

▷ Organizations using "systems drafting" would require little additional effort to develop a detail banking program. Systems drafting requires its users to organize, format, and standardize graphic information. These same standards fulfill many requirements identified for detail banking. Once the construction language is formalized, indexing details for storage becomes a simple task.

▷ System can be designed for manual, semi-automated, and automated retrieval of information.

▷ The descriptor system greatly increases the retrieval of relevant construction information.

▷ The searcher can specify the appropriate descriptors to define the limits of a search. Detail information can be retrieved in the following forms

1. actual details,
2. a numerical count of details associated with specific descriptors,
3. a printout of the detail numbers for a given search,
4. a Detail Data Card for each detail identified in the search,
5. a number of details using a specific descriptor term.

▷ Studies have shown this system to be the most direct and effective method of handling information in several other disciplines. Most in-house (individual) systems choose numbers, symbols, color coding, abbreviations and other complex methods to file information. Generally, these methods are only understood by the system originator. The recommended master system simply uses the English language for indexing details; it is as simple as associating subject area descriptors with a graphic picture.

▷ A great benefit to using this system is derived by indexing graphic information with more than one subject descriptor. Two or more descriptors for one graphic detail *greatly increases* its potential for retrieval, whereas filing a detail under one category or division of specifications limits its potential for retrieval.

▷ Multidescriptor indexing frees the information storage process from the category system. For example, if an indexer provides more than one descriptor to a detail, it will allow the researcher several avenues to retrieve the detail. However, if an indexer must select one category or division for filing detail, the searcher will likewise be forced to use only one category to retrieve detail. The later system presents a real problem when information can logically be filed under several categories.

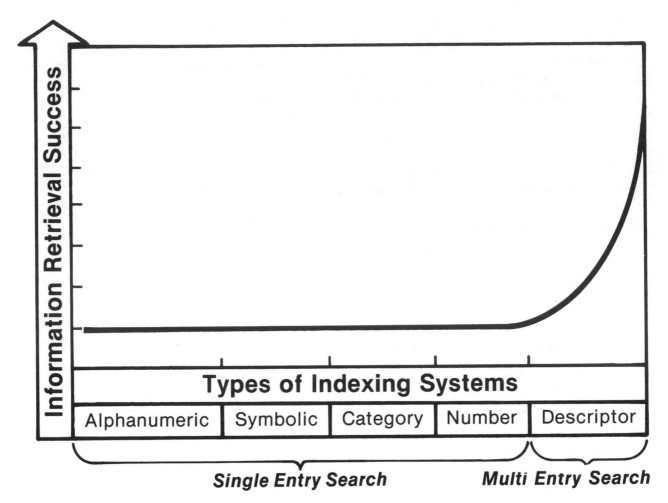

Figure 6-25   Information retrieval success: a comparison of systems.

▷ The system allows the searcher total freedom to prescribe the limits of a search by selecting appropriate descriptor terms, whereas a category system forces the indexer to decide where the information should be stored and under which categories it should be retrieved.

▷ Profile of retrieval success curve in Figure 6-25 is based on studies, user surveys, research, and actual use of systems. The complexity of most information-handling systems makes it difficult to tabulate the percentage of efficiency. Future studies should be conducted to obtain effectiveness ratios for each system. However, firsthand experience provides sufficient evidence that a descriptor system increases the potential to obtain relevant material because it is a *multientry search process*.

▷ The author's firsthand experience with a language, thesaurus, and descriptor system demonstrated the following benefits:

○ Effective indexing of information with limited training time required.

○ Search structured quickly with aid of thesaurus.

○ Rapid retrieval of relevant information stored in both a manual and computer system.

○ Ease in identifying closely related information.

○ Capability of pinpointing critical information from a large volume of material in minimum amount of time.

SEVEN
DEVELOPING A COMPREHENSIVE INFORMATION SYSTEM

7-1 HOW TO DEVELOP A COMPREHENSIVE INFORMATION SYSTEM

Structuring the detail banking program as part of a comprehensive information-handling system should be the goal of every design firm. The recommended system for handling details is valid and appropriate for processing all forms of information. Since the system is language dependent, it is critical that a comprehensive construction thesaurus be developed as part of the office procedures manual. *As pointed out earlier, the great value of this system is derived out of the development and use of a construction language that generates the base terminology for all construction documents.*

This system has the potential to bring about (1) a standardization of terminology, (2) consistency in document preparation, (3) a common language for working drawings, specifications, and information handling, and (4) improved coordination and communication throughout the construction industry. Expansion of the information-handling capabilities of this system requires the development of terminology in the nine classifications of subject areas discussed in Section 4-4.8.

The framework and requirements for a comprehensive information-handling system are similar and compatible to the procedures and guidelines set forth in detail banking. To expand the system simply requires expanding (1) the construction thesaurus and (2) the accession numbering system. An expanded system works with books, articles, papers, index cards, and abstracts just as the detail banking system works with details and data cards. For example, the Document Index Card and abstract are comparable to the Detail Data Card, while the actual book, article, paper, or building cost summary form are similar to the detail. The samples in Figure 7–1 through 7-4 (pages 158–161) demonstrate the similarity in materials processed for storage.

Successful information-handling experiences in other disciplines have the following common characteristics:

1. A standard language with descriptor terms structured in a thesaurus.
2. A standard format for processing information.
3. A system for document indexing and storing by accession numbers.
4. A Boolean-type search process linked to descriptor terms.

To develop a construction information center requires setting up a systematic method for collecting current information. Major sources of information must be contacted for automatic mailings and for special research reports. Some of the following sources should be considered when setting up the resource center and information-handling program:

▷ Construction information systems.
▷ Library centers.
▷ Building maintenance departments.
▷ Building research centers.
▷ Testing laboratories.

Example 1   *Document Index Card*

▷ University research centers.

▷ Professional societies and associations.

Testing laboratories and research stations currently doing material and product evaluations should be considered as a prime source for valuable performance information. A direct link to these agencies will help establish a source for current:

▷ evaluation data

▷ performance feedback

▷ documented problems

▷ test results.

Data collected in a comprehensive information-handling program will generate the base to develop functional details for the banking system. Many of the following benefits can be realized through a comprehensive construction information system:

▷ Less duplication in collecting, indexing, and storing construction information.

▷ Opportunities to recall valuable learning experiences.

▷ Less fragmentation in processing construction information.

▷ Costly errors prevented from oversights in using inadequate information.

▷ Less time required to convey essential information to design personnel.

▷ Uniformity and consistency in processing construction information.

▷ Less expensive to obtain required information in a limited time frame.

---

1 | EF 001 748

2 | *Title*   What's Happening to the Campus?
3 | *Author*   Brubaker, Charles William
4 | *Source*   The Perkins & Will Partnership
5 | *Imprint*   The Perkins & Will Partnership
    Architects
    Chicago, Illinois
    April 1968

6 | *ABSTRACT*

TRENDS IN CAMPUS PLANNING ARE DEVELOPED IN TERMS OF CHANGING EDUCATIONAL METHODS AND SOCIAL DEMANDS. MAJOR TOPICS COVERED ARE: (1) RE-EVALUATING THE NATURE OF LEARNING, (2) THE EFFECT OF TECHNOLOGY, (3) THE CAMPUS AS A COMMUNITY CULTURAL-EDUCATIONAL CENTER, (4) THE COLLEGE AND THE URBAN CRISIS, (5) THE MULTILOCATION COLLEGE, AND (6) EDUCATIONAL BUILDING SYSTEMS. WITHIN THESE CONTEXTS, DESIGN RECOMMENDATIONS AND SUGGESTIONS ARE OUTLINED, AND GRAPHIC EXAMPLES ARE PROVIDED. (MM)

7 | *DESCRIPTORS*

*BUILDING INNOVATION, *CAMPUS PLANNING, *DESIGN NEEDS, *EDUCATIONAL CHANGE, AUDIOVISUAL AIDS, COMMUNITY COLLEGES, COMMUNITY DEVELOPMENT, COMPUTERS, LEARNING PROCESS, SCHEDULING, STRUCTURAL BUILDING SYSTEMS, URBAN AREAS, URBAN UNIVERSITIES.

KEY

1. Accession number.
2. Title of document.
3. Personal author.
4. The institutional (source) is the corporate author or organizational source of a document and implies responsibility for issuing the document.
5. Document availability: place, address, and date of publication.
6. Abstract.
7. Descriptor terms indicating the major subject areas of the document.

Figure 7-1   Document EF 001 748 index card.

MOOER'S LAW

An information retrieval system will tend not to be used whenever it is more painful and troublesome for a customer to have information than for him not to have it.

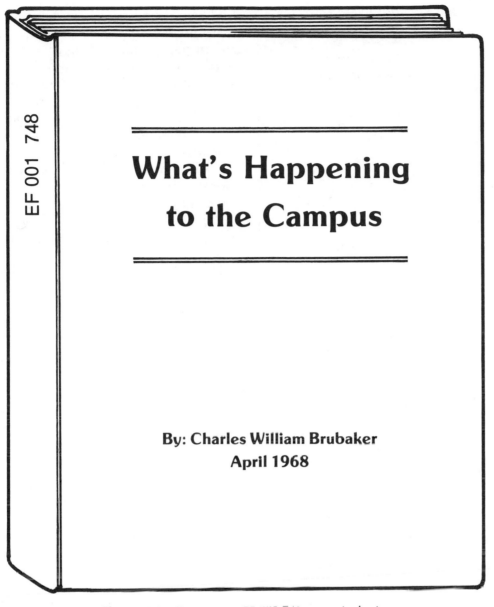

Figure 7-2 Document EF 001 748 cover indexing.

7-2 KEEPING PACE WITH A CHANGING COMMUNICATION TECHNOLOGY

Advancements in communication technology will enable new developments in detail banking systems. The next ten years may bring about the greatest change in word-processing and graphic communications equipment than ever before in history. In the 1980s the construction industry can expect to avail of improvements in a number of automation systems, including:

▷ word processors

▷ mini-computers

▷ computer-aided design

▷ computer data banks

▷ regional networks, and

▷ user terminals.

Example 2   Document Index Card

EF 002 328

Title   Bibliography of Environmental Design
           References
Author
Source   Wisconsin University, Environmental
           Design Center; ERIC Clearinghouse
           on Educational Facilities

Imprint: Environmental Design Center
           329 State Street
           Madison, Wisconsin 53703
           January 1968

ABSTRACT

A BIBLIOGRAPHY ON SOURCES RELATED
TO THE STRUCTURING OF THE PHYSICAL
ENVIRONMENT HAS BEEN DEVELOPED
BASED ON PROFESSIONAL AND CLASS-
ROOM EXPERIENCE. THIS INITIAL SELEC-
TION OF JOURNALS, BOOKS, AND UNPUB-
LISHED PAPERS GIVES AN OVERVIEW OF
PEOPLE AND THE ENVIRONMENTAL CON-
DITIONS WHICH ARE PART OF THEIR
DAILY LIVING PATTERN. INFORMATION
LEADING TO DESIGN PRINCIPLES AND
IMPLICATIONS BASED ON THE PHYSIO-
LOGICAL, PSYCHOLOGICAL, AND SOCIO-
LOGICAL NEEDS CAN BE SYNTHESIZED
FROM THE LITERATURE. MAJOR AREAS
OF COVERAGE INCLUDE: (1) ARCHITEC-
TURE, ENGINEERING, AND DESIGN, (2)
EDUCATION, HOUSING, AND URBAN
PLANNING, AND (3) ANTHROPOLOGY,
PSYCHOLOGY, PHYSIOLOGY, AND SOCI-
OLOGY. AN INTENSIVE PERIODICAL SUR-
VEY IS GIVEN FOR FORM PERCEPTION.
THE COVERAGE ALSO COVERS NUMER-
OUS SUBAREAS INTO GENERAL RANGE
OF DESIGN THEORY AND BEHAVIORAL
SCIENCE. A SELECTED SAMPLING OF EN-
VIRONMENTAL DESIGN STUDENT RE-
SEARCH REPORTS IS AVAILABLE FROM
THE CENTER. (MM)

DESCRIPTORS

*BEHAVIORAL SCIENCES; *BIOLOGICAL
SCIENCES; *ENVIRONMENTAL CRITERIA;
*ENVIRONMENTAL INFLUENCES; *PHYS-
ICAL ENVIRONMENT; ARCHITECTURE;
BUILDING MATERIALS; CITY PLANNING;
COLOR PLANNING; DESIGN; DESIGN
NEEDS; EDUCATIONAL FACILITIES; EN-
GINEERING; HOUSING; HUMAN ENGI-
NEERING; LIGHTING; PERCEPTION;
PHYSIOLOGY; PSYCHOLOGICAL STUD-
IES; SOCIOLOGY

Present availability-ERIC System.

Figure 7-3   Document EF 002 328 index card.

These new systems make it possible to develop a design station capable of retrieving valuable information from central data banks. Use of automation equipment will make instantaneous retrieval of useful information available for the design decision-making process. All stored information will be retrieved, manipulated, and restructured for graphic communication in a moment's notice.

Communication demands by design professionals can encourage developments in word processing and other types of graphic display equipment. Word processors will soon have greater capabilities to retrieve information using the *Boolean retrieval approach.* This improvement will enable the researcher to work with a multi-descriptor search from a large data base at a lower system cost. Within the near future it will be possible to integrate word processors with micro-film processing and other storage equipment. These machines will be able to contain the actual film for each detail stored in the system. By connecting systems, the user will be able to select appropriate details by descriptor input to the word processor which in turn will activate a search for the desired detail in the microfilm storage equipment (Figure 7-5).

Design professionals should research and evaluate existing and proposed communication systems so that in-house information-handling programs can be formatted to be compatible with automation and future systems. *Communication system projections will enable the designer to stage information system development progressively as opposed to terminating one system and creating a new one each time communication technology changes.*

It is important to develop a system that is flexible and expandable so that new areas of information can be easily stored within the existing system. A system that will handle only predetermined categories of data makes it impossible to expand the system without changing the overall framework. These systems are very costly and short-lived.

7-3   PLANNING FOR CHANGE

Information system success can be measured by performance during periods of great change in construction and communication technology. A

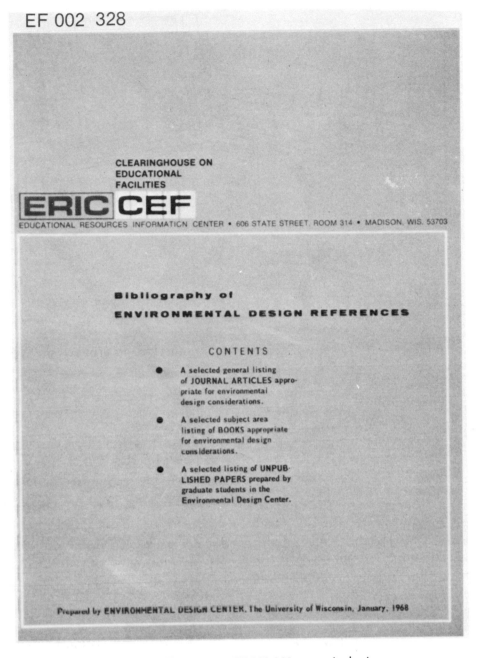

Figure 7-4 Document EF 002 328 cover indexing.

system designed with flexibility and adaptability to technology change will provide the greatest benefits to its users. The following directives will help system designers plan for change:

1. Develop in-house information programs with flexibility and in accordance with established standards.

2. Organize information-handling systems into a standard format

○ construction language
○ detail terminology
○ coding and indexing systems

3. Use subject area descriptors and numbering systems that are compatible with communication and information-handling systems.

4. Select equipment that is capable of linking into new systems without changing storage and retrieval programs.

Figure 7-5 Developing technological proficiency in production.

5. Most important, plan and program information-handling activities before renting or purchasing equipment. (Remember: *Time is money. Planning is a must.*)

New computer-aided drafting systems are now entering the market. These systems are equipped with cathode ray tubes and keyboards along with digitizers, plotters, and printers capable of providing instantaneous graphic responses. With this type of equipment the designer will be able to bank and retrieve details for immediate review, evaluation, and preparation on the final working drawings. Only three major steps are required to process details for computer-aided production:

1. Store details in computer Detail Bank, numbering details in the sequence of development.
2. Enter Detail Data Card information by descriptor and accession number.
3. Retrieve details by descriptor terms, using single or multidescriptors.

Access to many types of field-evaluated and approved details will be a tremendous asset to the draftspersons preparing working drawings. A detail banking system will provide the source for:

1. Making early project development cost and time projections.
2. Preparing preliminary design proposals.
3. Developing mock-up details and drawing sheets.
4. Selecting appropriate details.
5. Structuring new details.

To build a detail bank, five major sources should be used:

1. **Details from Existing Drawings.** A search of existing drawings will provide an excellent source for repetitive details that have been used successfully. These details should be evaluated and considered for input to the computer bank.
2. **Details from Developments of New Projects.** Details will emerge from the requirements and needs established for new projects. As these details are developed, evaluated, and approved they can be processed for detail banking.

3. **Manufacturer's Product Details.** Details from selected manufacturers' products can be used in detail banking. Product catalogs can provide the source for many reusable details.

4. **Photographed Construction Details.** Following construction projects with a camera can provide the source for excellent details. The photographic methods of detail banking can save time and money in preparing working drawings.

5. **Detail Clearinghouses.** Collections of approved details are beginning to emerge in different service bureaus across the nation. These newly developed centers are begin-

ning to provide standard details to other design professionals for a user fee. Detail bankers should watch for the development of new detail clearinghouses.

A computer-aided banking system will provide the designer with a maximum amount of freedom to restructure details for instant reuse. By automating detail banks, the designer will be provided with a wide range of design options as well as total flexibility to redesign and alter standard or typical details. Figure 7-6 demonstrates detail processing for a computer-aided banking and production system.

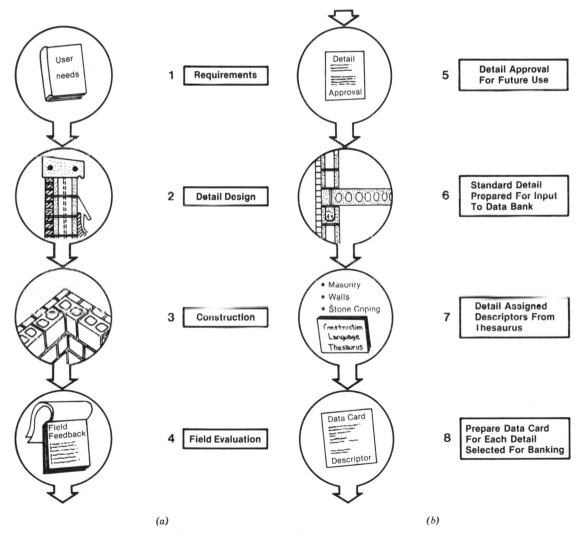

(a)   (b)

Figure 7-6 Banking construction details using a computer system.

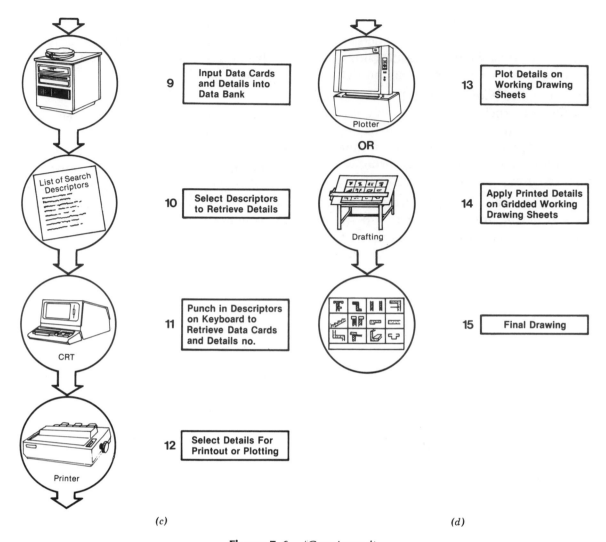

9 Input Data Cards and Details into Data Bank

10 Select Descriptors to Retrieve Details

11 Punch in Descriptors on Keyboard to Retrieve Data Cards and Details no.

12 Select Details For Printout or Plotting

13 Plot Details on Working Drawing Sheets

14 Apply Printed Details on Gridded Working Drawing Sheets

15 Final Drawing

(c)   (d)

Figure 7-6 (Continued)

Both computers and word processors can store effectively specifications and text information for later printout and use on graphic drawings. This capability provides the designer with an opportunity to structure and edit working drawings schedules on a screen before final printout. Selected letter styles also enables the designer to have complete flexibility in preparing standardized text for all details and working drawings.

Automation systems can provide many opportunities to store different types of details, schedules, and performance records for instantaneous retrieval by the draftsperson. These advancements in communication technology are continuing to lower overall labor costs by reducing the time required to produce high-quality con-

struction documents. Properly coordinated systems will benefit the design professional by allowing more time for design development stages.

As a greater number of reusable details are banked, it will become easier to exchange information. This concept can provide additional savings in developing working drawings. Centralized detail banks accessible through computer will make it possible to share information instantaneously. This will enable the design professional to reduce repetitive activities associated with detail development. Field-evaluated details can be screened and approved for the detail banking system. By making use of evaluated design solutions, it will be possible to reduce risks

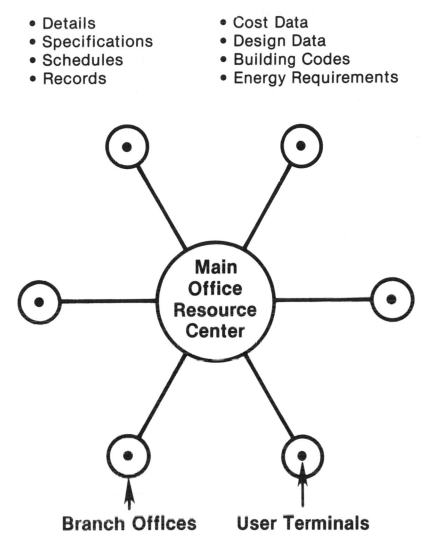

**Transferring Construction Information
From Central Data Bank**

- Details
- Specifications
- Schedules
- Records

- Cost Data
- Design Data
- Building Codes
- Energy Requirements

Branch Offices   **User Terminals**

Figure 7-7   Linking multiuser terminals in branch offices.

associated with detail development. This could lead to reduced liability insurance and time lost settling legal suits arising out of detail problems.

Larger firms or departments can make greater use of automation by creating user terminals in all branch offices. Direct line ties back to a central data bank will enable users to access information quickly and easily without costly time delays. A network of user terminals can increase the cost effectiveness of the system by allowing a greater number of users to access a resource center (Figure 7-7).

7-4 DEVELOPING A NETWORK OF INFORMATION EXCHANGE

Information flow is critical to the design process. To cope with rapidly increasing volumes of construction information, it is necessary to coordinate centralized processing efforts. Major steps must be taken to reduce duplication of information collection and processing activities within each organization. A network for information exchange can be supported for the following reasons.

▷ To reduce repetitive development and evaluation activities from one office or department to another.

▷ To speed up the design, evaluation, selection, and approval process.

▷ To develop uniformity in the design and evaluation process.

▷ To reduce the overall cost in developing and evaluating construction details.

▷ To give all size departments the opportunity to access the same information.

▷ To take full advantage of establishing a detail banking system.

With many economic and legal constraints facing the design professions, it is important to consider new systems in construction detailing. Detail banking can provide organizations with new alternatives in working drawing development while realizing direct benefits in (1) time savings, (2) cost savings, and (3) quality control. The decision-making process can also be enhanced by access to the historic files of evaluated and approved construction details. By detail banking, each firm can achieve a higher quality of graphic communication in working drawing production. This can ultimately lead to improved design and increased performance of materials, products, and facilities.

GLOSSARY

APERTURE CARD A card with a rectangular hole or holes specifically prepared for the mounting or insertion of microfilm.

DESCRIPTORS Derivatives of expressions or terms. Any single term or multiword term may be used for indexing a document but should reflect the language in the literature, have an agreed-upon meaning, and arise frequently in the document. Descriptor systems have limited vocabularies. A descriptor dictionary is usually maintained, with "scope notes" to define the scope and meaning of each descriptor for the system.

KEY WORD A significant or informative word in a title, abstract, body, or part of the text generally used to describe a document.

REPETITIVE DETAIL A detail that recurs frequently in the design process of a construction project. It can be referred to several times within a set of working drawings or in a series of contract documents.

REUSABLE DETAIL A detail that has the potential qualities to be used again or repeatedly in the construction process.

STANDARD DETAIL A detail that has been evaluated as an established base of agreed upon quality and acceptability for application in construction. It has been recognized as having lasting value for future use or reference.

SUBJECT AREA INDEXING A system based on a broad grouping of related terms and/or concepts. A distinction is made between indirect and direct entry, or word and controlled indexing. In an indirect entry system a common heading is used to guide an alphabetical sequence by qualifying inversions. Example: Glass, Plate. An entry that is listed as "Plate Glass" is an example of the direct entry of a subject area index.

THESAURUS A controlled authority list of terms displaying semantic and conceptual relationships between terms. It is a collection of words or information pertaining to a particular field or set of concepts. A thesaurus is generally used as a word reminder list. The words used to identify information are defined in a connative manner, so two words found in the same document will reflect on the meaning of each other. The vocabulary is limited and detailed for the search.

REFERENCES

1. Oklahoma Chapter American Institute of Architects. *AIA News Letter.* September 1977.

2. Varney, Sexton and Sydnor. "Architects' Systematic Use of Standard Detail Library." *The Paper Plane,* vol. 2, no. 7. June 1979. MRH Associates, Inc., P.O. Box 11316, Newington, CT 06111.

3. *Manual of Facts on Concrete Masonry (A).* National Concrete Masonry Association, P.O. Box 781, Herndon, VA 22070.

4. *Roofing Manual.* National Roofing Contractors Association, 221 North LaSalle Street, Chicago, IL 60601.

5. "Quality Control in the Preparation of Working Drawings." *Guidelines for Improving Practice—Architects and Engineers Professional Liability.* Victor O. Schinnerer & Company in Cooperation with American Institute of Architects, the National Society of Professional Engineers/Professional Engineers in Private Practice, and CNA/Insurance, 1971.

6. William J. C Connell. *Graphic Communication in Architecture.* Champaign, IL: Stipes Publishing Company, 1972.

7 Northern California Chapter American Institute of Architects. *The POP Manual: Recommended Standards on Production Procedures,* vols. 1–3. Committee on Production Office Procedures, The American Institute of Architects, 790 Market Street, San Francisco, CA 94102, July, 1980.

8. Philip M. Bennett and Judy A. Jones. *Construction Information Systems Study.* The University of Wisconsin—Extension. Department of Engineering and Applied Science. For the Construction Sciences Research Foundation, Washington, D.C. © 1978.

9. *ERIC. Rules for Thesaurus Preparation.* Office of Education, Panel on Educational Terminology (PET). U.S. Office of Education, U.S. Department of Health, Education and Welfare. Superintendent of Documents Catalog No. FS5.212: 12047. U.S. Government Printing Office, Washington, D.C. 20402, 1969.

10. *Uniform Construction Index: A System of Formats for Specifications, Data Filing, Cost Analysis and Project Filing.* The Construction Specifications Institute, 601 Madison Street, Alexandria, VA 22314.

11. *Canadian Thesaurus of Construction Science and Technology (TCCS),* 2 vols. Available from J. L. Hall, Construction Division, Construction and Consulting Services Branch, Industry, Trade and Commerce, 235 Queen Street, 7th Floor E., Ottawa, Ontario, K1A0H5, July 1978.

12. *Rules for Preparing and Updating Engineering Thesauri.* Engineers Joint Council, June, 1965. Available from Publications Department, American Association of Engineering Societies, 345 East 47th Street, New York, NY 10017.

13. ERIC Processing and Reference Facility. *Thesaurus of ERIC Descriptors.* Working Copy, Descriptor Listing. June 1970 and latest editions. Operated for U.S. Office of Education by LEASCO Systems and Research Corporation. Bethesda, MD.

14. *Construction Industry Thesaurus.* 2nd ed. July 1976, Department of the Environment. CIT Agency at the Polytechnic of the South Bank, Wandsworth Road, London SW82JZ.

15. *ERIC. Thesaurus of ERIC Descriptors.* Completely Revised 1980. U.S. Office of Educa-

tion. Onyx Press, 2214 North Central Avenue at Encanto, Phoenix, AZ 85004.

16. *Thesaurus of Engineering and Scientific Terminology*. American Association of Engineering Societies (formerly Joint Engineers Council). Publications Department, 345 East 47th Street, New York, NY 10017.

17. District of Columbia Metropolitan Chapter, The Construction Specifications Institute. *A Glossary of Construction Specifications Terminology*. 1777 Church Street, N.W., Washington, D.C. 20036.

18. *CIT Definitions of Construction Terms*. Construction Industry Thesaurus. Polytechnic of the South Bank, Wandsworth Road, London SW8 2JZ, 1981.

19. *HUD Research Thesaurus*. U.S. Department of Housing and Urban Development, Office of Policy Development and Research. Superintendent of Documents. U.S. Government Printing Office, Washington, D.C. 20402, 1980.

20. *Manual for Building a Technical Thesaurus*. No. AD633279. National Technical Information Service (NTIS), Department of Commerce, 5285 Port Royal Road, Springfield, VA 22161.

21. Philip M. Bennett. "How to Develop a Detail Banking System," *The Paper Plane*, Volume III, Number 6, May 1980. MRH Associates, Inc., P.O Box 11316, Newington, CT 06111.

FURTHER READING

American National Standard Nomenclature for Steel Doors and Steel Door Frames. No. ANAI-A123.1. Latest Edition. Steel Door Institute, Cleveland, OH.

Bennett, Philip M. "Are You Banking Your Construction Details?" *The Paper Plane,* April, 1980. MRH Associates, Inc., P.O. Box 11316, Newington, CT 06111.

Brooks, Hugh. *Illustrated Encyclopedic Dictionary of Building and Construction Terms.* Englewood Cliffs, NJ: Prentice-Hall, 1976.

Building Failures Forum. Editor/Publisher, Raymond A. DiPasquale. P.O. Box 848, Ithaca, NY 14850.

DDC Retrieval and Indexing Terminology. No. ADA068500. National Technical Information Services (NTIS), Department of Commerce, 5285 Port Royal Road, Springfield, VA 22162.

Directory of Federally Supported Information Analysis Centers. 4th ed. National Referral Center, Library of Congress, Washington. Superintendent of Documents, U.S. Government Printing Office, Washington, D.C., 20402. Stock No. 030-000-00115-0, 1979.

Glossary of Architectural Metal Terms. The National Association of Architectural Metal Manufacturers, Suite 2026, 221 North La-Salle Street, Chicago, IL 60601.

Greater Phoenix, Arizona Chapter #98 of the National Association of Women in Construction. *Construction Dictionary,* P.O. Box 6142, Phoenix, AZ 85005 1978.

Harris, Cyril M. *Dictionary of Architecture and Construction.* New York: McGraw-Hill, 1975.

Information Retrieval Systems. General Services Administration, National Archives and Records Service Office of Records Management, Federal Stock No. 7610-181-7577. Superintendent of Documents, U.S. Government Printing Office, Washington, D.C. 20402, 1970.

International Masonry Institute. *The Masonry Glossary.* Masonry Institute of Wisconsin, 4300 W. Brown Deer Road, Milwaukee, Wis.

Jarvis, Donald E. *Integraphics: An Experiment in Architectural Communication.* Jarvis Putty Jarvis, Inc., 2010 One Main Place, Dallas, TX 75250, April 1973.

Liebing, Ralph W., and Mimi Ford Paul. *Architectural Working Drawings,* 2nd ed., New York, Wiley Inc., 1983.

MASTERFORMAT: Master List of Section Titles and Numbers. CSI Document MP-2-1. The Construction Specifications Institute, 601 Madison Street, Alexandria, VA 22314.

Newman, Morton. *Standard Structural Details for Building Construction.* New York: McGraw-Hill, 1968.

Paper Plane (The). Editor, George S. Borkovich. MRH Associates, Inc., P.O. Box 11316, Newington, CT 06111.

Powers, Edgar, Jr., Editor. *Systems Drafting Manual for Architectural & Engineering Firms.* Gresham, Smith and Partners, 3310 West End Avenue, Nashville, TN 37203, 1980.

Powers, Edgar, Jr., Editor. *Systems Drafting Manual for Reprographic Firms,* Gresham, Smith and Partners, 3310 West End Avenue, Nashville, TN 37203, 1979.

Powers, Edgar, Jr. *UNIGRAFS—Unique Graphics for Architects and Engineers.* Gresham, Smith and Partners, 3310 West End Avenue, Nashville, TN 37203, 1981.

Rosen, Harold J., and Philip M. Bennett. *Construction Materials Evaluation & Selection—A Systematic Approach. New York: Wiley, 1979.

Stein, J. Stewart, *Construction Glossary: An Encyclopedic Reference and Manual*. New York: Wiley, 1980.

Stitt, Fred. A. *Systems Drafting: Creative Reprographics for Architects and Engineers*. New York: McGraw-Hill, 1980.

Wakita, Osamu, and Richard Linde. *The Professional Practice of Architectural Detailing*. New York: Wiley, 1977.

INDEX

Abbreviations, 65, 67, 83
Accession numbers, 133, 142, 143, 150
Acronyms, 65, 67, 83
AIA, 68, 73, 91
Alphanumeric categories, 149
Alphanumeric systems, 107, 113
  advantages and disadvantages, 113
Aperture cards, 133, 167
Automated systems, 133, 134, 136, 138, 139

Barriers, 64, 65
Boolean retrieval, 135, 153, 154, 157, 160
Building failures, 24, 25, 52
Building materials:
  evaluation, 9, 12, 13, 24
  selection, 24

Canada, 1, 2, 95
Canadian Thesaurus of Construction Science and Technology, 68, 76, 82, 91, 94, 115
  applications, 121, 124, 125, 128
Checklists, 47, 48
Clearinghouse on Educational Facilities, 68, 83
Color coding, 14, 15
Committees:
  advisory, 73
  control, 77
Communication, 5, 23-25, 28, 48, 52, 55, 63
Communication technology, 159-161
Comprehensive Construction Information System, 157
Computers, 47, 63, 68, 159
  computer-aided drafting, 162-164
  information storage and retrieval, 134, 136, 138, 139
Construction claims, 23, 25
Construction details, 1, 3, 5, 23
Construction industry, 1, 2, 7, 63, 64, 65, 68, 73
Construction Industry Thesaurus, 76, 80, 82, 91, 93, 100, 115, 125, 128
Construction information, 1, 3, 7, 30, 59, 68
  categories, 67
  cross-reference, 65, 76, 80, 82, 152, 153
  indexing, 67, 73, 80, 82, 94-114, 128, 129, 147, 149, 150, 151
  storage and retrieval, 1, 2, 4, 5, 59, 64, 68, 82, 129, 132
  subject area indexing, 167
Construction information center, 157, 164, 165
Construction thesaurus, 23, 42
Controlled vocabulary, 154
Criteria selection, 5, 7
C S I, 68, 73, 91, 95, 129
C S I Uniform Construction Index, 68, 76, 90, 93, 95

Data bank, 1, 5
Descriptor terms, 69, 80-91, 94, 115-118, 128, 129, 132, 151-154, 167
  broader terms, 80, 82, 117, 152
  cross-reference, 80, 152, 153
  homograph, 117
  multiword, 116, 121, 153, 155, 160
  narrower terms, 80, 82, 117, 152
  parenthetical qualifiers, 117
  related terms, 82, 117, 152
  scope notes, 82, 117
  selection, 118-132
  singleword, 115, 116, 121
  structure, 118
  used for terms, 82, 117
Descriptor Term System, 94, 114, 115, 129
  advantages and disadvantages, 114, 128, 129
  development, 115
Detail banking, 1, 2, 5, 7, 33, 61, 73, 138, 166
  critical steps, 119-132
  multidisciplinary use, 124, 128, 130, 131
Detail Banking System, 3, 4, 5, 14, 47, 52, 55, 73, 80, 93, 94
  centralized, 164, 165
Detail clearinghouses, 163
Detail cycle, 61
Detail Data Bank, 2, 4, 5, 9, 162-165
Detail Data Card, 14, 52, 59, 60, 133, 135-138, 150, 162
  examples, 123, 127, 131
Detail filing, 95, 105
Detail performance requirements, 48
Detail preparation, 33, 42
Detail retrieval, 129, 133-139, 141, 151-154
Details, standard, 3, 4, 7-11, 18-24, 28, 30, 33, 167
  abstract, 24
  architectural and engineering, 18, 21, 22
  associations, 15, 19, 20
  categories, 7, 9, 14
  construction, 3, 5, 25, 26, 52, 57
  cross-referencing, 152, 153
  development, 5, 25, 28, 49
  grading, 9, 13, 24
  indexing and storing, 128, 129, 147-151
  manufacturers, 15-17, 163
  photographic, 15, 18, 163
  processing, 138, 139, 142
  realism, 24
  repetitive, 167
  reusable, 7, 14, 18, 26, 167
  selection, 9, 14
  standardized, 5, 7, 9, 14, 28, 30
Dewey Decimal System, 113
  disadvantages, 113

Document Index Card, 157, 158, 160

English language, 1, 73, 76
ERIC Thesaurus of Descriptors, 68, 76, 82–90, 115, 151

Field evaluation, 52, 55–57, 59
Field Evaluation Data Card, 55, 56
Field feedback, 1, 4, 5, 23, 52, 55–59

Generic terms, 24
Glossary of Construction Specifications Terminology, 79
Graph, frequency, 14
    detail reuse, 14

Historic file, 5, 7, 9, 14, 23, 59, 73, 138
HUD Research Thesaurus, 91
Human memory, 71

Information exchange:
    network, 165, 166
    reasons for, 166
Information handling, 1–3, 63–65, 67, 68, 94, 154
    classification system, 93, 105, 112
    system evaluation, 93, 94, 112–114
Information search process, 65, 66, 73, 74
Inhouse systems, 65, 73, 76, 91

Joint Council of Engineers. 1. 69. 82, 115

Keyword, 95, 99, 129, 132, 153, 167

Language, 1, 2, 23, 24, 47, 90, 91
    common, 63, 64, 67, 68, 75, 80
    construction, 2, 23, 24, 63–69, 73, 76, 78, 83–90, 115
    preparation, 91
Legal problems, 23, 63
Log book, 57

Management programs, 49
Manual systems, 133–139
MASTERFORMAT, 96–99, 129
Master System, 1, 2, 107, 154–156
    reasons for, 154–156
Material Evaluation Form, 9, 12
Mini-thesaurus, 80–90, 119
Mock-up drawings, 33, 42–46
Moorer's Law, 158

Notation standard, 42, 47
NSPE, 91
Numbering systems, 142–151
    advantages and disadvantages, 147

analysis, 142–144
    components, 142–144
    level one, 142–151
    level two, 142–151
    selection, 144, 147

Office library, 23
Optical Coincidence Cards, 133

Quality control, 1, 5, 9, 18. 19, 47–49, 55

Research stations, 157, 158
Resources, library, 59
Retrieval descriptors, detail examples, 122, 123, 126, 127, 130, 131
Rules For Thesaurus Preparation, 82, 115–119

Search process, 151–154
    broad search, 151–154
    narrow search, 151–154
Semi-automated systems, 133–136, 139
Specifications, 23, 42, 47, 49, 63, 69, 73, 75, 95
    retrieval, 129, 132
Specification systems, 95–104, 113, 129, 132
    advantages and disadvantages, 113
Standardization, 1, 19, 33, 42, 47, 48, 64, 68, 132, 157
Standards, 33, 48, 75, 77, 78, 83
    office manual, 33, 42, 49, 79, 80, 81
Symbolism, 1, 65, 67
Synonyms, 65, 67, 73, 81, 83
Systematic method, 157
Systematic process, 5, 25, 26, 48, 57

Terminology, 1, 2, 23, 24, 63–65, 68, 69, 73, 75–81
    classifications of, 80
    construction, 1, 68, 81
    control, 47, 64, 67, 81
    families of, 80
    generic terms, 91, 152–154
    repetitive, 73, 76
    standardized, 65, 68, 75, 76, 132
Testing laboratories, 157, 158
Thesaurus, 2, 68–70, 76, 80–92, 100, 115, 167
    construction, 151
    preparation, 91, 115–118
Thesaurus of Engineering and Scientific Terminology, 76, 91

United States, 1, 2, 68, 95
Universal system, 107

Word processors, 47, 49, 63, 68, 133–136, 159, 160, 164
Working drawings, 69, 73, 164